OXFORD BIOGEOGRAPHY SERIES

Editors: A. HALLAM, B. R. ROSEN, AND T. C. WHITMORE

OXFORD BIOGEOGRAPHY SERIES

Editors

A. Hallam, School of Earth Sciences, University of Birmingham
B. R. Rosen, Department of Palaeontology, The Natural History Museum, London
T. C. Whitmore, Department of Geography, University of Cambridge

The aim of the series is to publish a range of titles demonstrating the breadth of biogeography, from the biological to the geological ends of the spectrum. The subject is being revolutionized by plate tectonics, molecular phylogeny, population models, vicariance, cladistics, and spatial classification analyses. For both specialist and non-specialist, the Oxford Biogeography Series will provide dynamic syntheses of new developments.

1. Whitmore, T. C. (ed.) *Wallace's line and plate tectonics*
2. Humphries, C. J. and Parenti, L. R. *Cladistic biogeography*
3. Whitmore, T. C. and Prance, G. T. (ed.) *Biogeography and Quaternary history in tropical America*
4. Whitmore, T. C. (ed.) *Biogeographical evolution of the Malay archipelago*
5. Aiken, S. R. and Leigh, C. H. *Vanishing rain forests: the ecological transition in Malaysia*
6. Adam, P. *Australian rainforests*
7. George, W. and Lavocat, R. *The Africa–South America connection*
8. Yin Hongfu (ed.) *The palaeobiogeography of China*
9. Kohn, A. J. and Perron, F. E. *Life history and biogeography: patterns in Conus*
10. Hallam, A. *An outline of Phanerozoic biogeography*

An Outline of Phanerozoic Biogeography

ANTHONY HALLAM

*Lapworth Professor of Geology in
the University of Birmingham*

OXFORD   NEW YORK   TOKYO
OXFORD UNIVERSITY PRESS
1994

Oxford University Press, Walton Street, Oxford OX2 6DP

Oxford   New York
Athens   Auckland   Bangkok   Bombay
Calcutta   Cape Town   Dar es Salaam   Delhi
Florence   Hong Kong   Istanbul   Karachi
Kuala Lumpur   Madras   Madrid   Melbourne
Mexico City   Nairobi   Paris   Singapore
Taipei   Tokyo   Toronto
and associated companies in
Berlin   Ibadan

Oxford is a trade mark of Oxford University Press

Published in the United States
by Oxford University Press Inc., New York

A catalogue record for this book is available from the British Library

Library of Congress Cataloging in Publication Data
(Data available)

ISBN 0 19 854061 2 (Hbk)
ISBN 0 19 854060 4 (Pbk)

Typeset by The Electronic Book Factory Ltd, Fife
Printed in Great Britain by
Bookcraft (Bath) Ltd
Midsomer Norton, Avon

PREFACE

Since the 1970s the general acceptance of plate tectonics, implying continental drift, has proved a great stimulus to the study of the biogeography of the period after metazoan life first became established. Information is widely scattered in the literature, and it may be difficult to discern the general conclusions. This book represents the first attempt to synthesize the results of over two decades of research in a way which will be intelligible to a wide readership. Despite the immense amount of study neither rigorous analysis nor thorough global coverage of more than a limited number of fossil groups has been achieved, and any effort to write a comprehensive treatise would be premature. Accordingly what is attempted here must be treated as no more than an outline, a concise work intended to conform to the requirements of the Oxford Biogeography Series.

The challenge is to discern what general patterns may be inferred for different times of geological history, and to propose explanations for these patterns. This could help to 'set the agenda' for future research. Inevitably the discernible large-scale biogeographic patterns differ considerably through time, with the respective roles of continental relationships, climate, and sea level changing more or less continuously. The book is aimed not just at palaeontologists involved with biogeography but also those geologists and biologists with relevant interests. Geologists may find it a useful source of information on the great value that fossils can provide for tectonic reconstructions, while biogeographers may benefit from the historical perspective provided by taking several hundred million years of Earth history into account.

No individual subject is treated at length, but the comprehensive bibliography should provide a more than adequate introduction to a wide-ranging literature for those who wish to pursue matters in more depth. Biologists, and perhaps even some geologists, should welcome the inclusion of the Phanerozoic time scale as an appendix. An attempt has been made to minimize the use of jargon to explain simple concepts. The illustrations are not original. Where they are reproduced exactly as they were first published, permission has been sought from the authors and publishers concerned. Where I have significantly altered any, I have acknowledged this and signalled the change by such phrases as *adapted from* or *simplified from*.

In my understanding of palaeobiogeography I have benefited over the years from stimulating discussion with many colleagues, among whom I would single out Bill Chaloner, Richard Fortey, Malcolm McKenna, Brian Rosen, Tim Whitmore, and Fred Ziegler. I am also indebted to Lorna Viikna and June Andrews for the typing, and Jackie Stokes and Liz Smith for drafting the diagrams.

Birmingham A. H.
November 1993

CONTENTS

1

Introduction

HISTORICAL DEVELOPMENT OF BIOGEOGRAPHY

The origin of biogeography as a scientific discipline has been attributed to the great French naturalist of the Enlightenment, Buffon. In 1761 he pointed out that the Old and New Worlds have no mammalian species in common. This led to a generalization sometimes known as Buffon's Law, that different regions of the globe, though experiencing the same environmental conditions, were inhabited by different species of animal and plant. Such a 'law' implies that the key factor controlling organic distribution is history rather than ecology. It was actually the French botanist Candolle who was the first to make this generalization, early in the nineteenth century, and his 1820 publication can be taken as marking the true birth of biogeography (Nelson 1978). In this work Candolle lists twenty botanical regions across the world, defined by endemic species.

Despite Nelson's promotion of Candolle, Brundin (1988) considers that another botanist, Joseph Hooker (1853), was the real founder of causal historical biogeography. From the results of his comprehensive taxonomic analysis of southern floras, he inferred that those at the southern tip of South America, New Zealand, and southern Australia were closely related. As a result of investigating possible dispersal mechanisms he became convinced that dispersal was not a significant agent and concluded that the southern continents must once have been connected – an anticipation of Suess's concept of Gondwanaland, put forward at the end of the nineteenth century. Hooker's explanation was based on vicariance biogeography, thus anticipating a school of thought that has arisen within the last two decades.

Hooker's contemporary Charles Darwin was a pioneer of dispersalist biogeography. Chapters 12 and 13 of *The Origin of Species* (1859) are devoted to geographic distribution. He concluded that 'all the grand leading facts of geographic distribution are explicable on the theory of migration, together with subsequent modification and the multiplication of new forms. We can thus understand the high importance of barriers, whether of land or water, in not only separating, but in apparently forming the several zoological and botanical provinces'. Alfred Russell Wallace (1876) expanded and developed these chapters and produced his famous classification into six zoogeographic regions of the continents based on vertebrates and land molluscs: Palaearctic; Ethiopian; Oriental; Australian; Nearctic; and Neotropic. Sclater (1858) had earlier produced a very similar classification based on birds.

The dispersalist interpretation, implying migrations of species from 'centres of origin', remained the dominant one until the 1970s, and is associated in particular with the publications of Matthew (1915), Darlington (1957), and Simpson (1943, 1947).

The difficulties that this dispersalist approach could lead to, accepting a stabilist view of the continents, is exemplified by a paper by Darlington (1948) on the distribution of cold-blooded vertebrates. He pointed out that certain primitive and, by general consent, closely related groups are found only in Africa and South America. These include pelomedusid tortoises (also known in Madagascar), pipid frogs, and several families of freshwater fish including characins, osteoglossids, and cyprinodonts. Darlington recognized that there are only three ways of accounting for this distribution: the various organisms were able to migrate directly across the Atlantic Ocean, by a South Atlantic land connection or by a long land journey via Asia and North America. The first possibility was considered highly implausible and the second rejected because it departed from orthodox geological views. Darlington was left with the hypothesis of a migration across the Holarctic Realm which has left no trace in the indigenous fauna, living or fossil. The connection was presumed to have been during the Cretaceous.

Directly as a consequence of the general acceptance by geologists of continental drift, the alternative interpretation was put forward that formerly widespread biota have been split up (vicariated) by the establishment of barriers, such as newly opened oceans. The subject of vicariance biogeography will be dealt with more fully in Chapter 3.

Attention has often been restricted to the distribution of living organisms, because many biogeographers paid no attention to the fossil record. By the end of the last century it had become apparent to palaeontologists that the close taxonomic relationships of Mesozoic terrestrial vertebrates and other groups between South America, Africa, and India implied very different world geography in that era (Hallam 1967*a*). The conventional interpretation was in the form of continental land bridges that had subsequently foundered beneath the South Atlantic and Indian Oceans (Fig. 1.1). Alfred Wegener (1924) cited this biogeographic evidence in support of his hypothesis of continental drift, pointing out that neither the geological evidence (absence of granitic rocks) nor the geophysical evidence (high density of ocean floor) supported the idea of foundered continents, and that the only plausible alternative was that the Atlantic and Indian Oceans had opened in the fairly recent geological past. This interpretation did not convert the palaeontologists any more than it did the

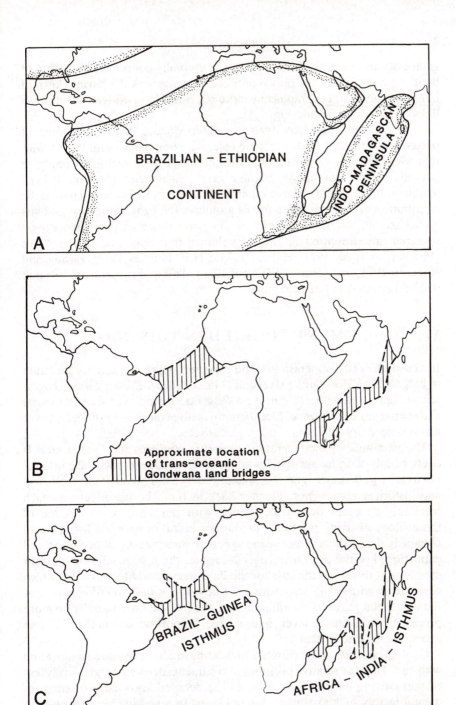

Fig. 1.1 Trans-oceanic connections between South America, Africa, and India according to (A) Neumayr (1887), (B) Schuchert (1932), and (C) Willis (1932). After Hallam (1967a)

geologists and geophysicists, despite the mutually contradictory nature of their arguments, and the only major consequence was the paring down of huge continents to geographically more restricted land bridges or 'isthmian links' (Hallam 1973*a* and Fig. 1.1).

It was not until the late 1960s, as a consequence of a vast amount of palaeomagnetic and oceanographic research after the Second World War, that Wegener was vindicated and the theory of plate tectonics, implying laterally mobile continents, became firmly established (Hallam 1973*a*). Within only a few years a large number of palaeontologists re-examined the distribution of their favoured fossils groups in the light of the new mobilist global tectonics, and the results of this early phase of palaeobiographic research are presented in several volumes that appeared in the 1970s (Middlemiss *et al.* 1971; Hallam 1973*b*; N.F. Hughes 1973; Tarling and Runcorn 1973; and Gray and Boucot 1979). This and subsequent research will be reviewed in later chapters.

THE TREATMENT ADOPTED IN THIS BOOK

It is customary to distinguish two end-members within the activity spectrum of biogeographic research (Myers and Giller 1988*a*). *Ecological biogeography* is concerned with ecological processes occurring over short temporal and spatial scales, *historical biogeography* with evolutionary processes over millions of years on a large, often global scale.

The vast bulk of biogeographic literature concerns the Quaternary. It deals mainly with extant species whose ecological tolerances are often if not usually well known, and is characterized by a fair blend of the ecological and historical approaches. Further back in time the historical approach inevitably becomes dominant, though with reference to knowledge of the ecology of living relatives. Taxonomic affinities with the living world obviously diminish with increasing age, as a consequence of evolution and extinction. For the pre-Quaternary Neogene, there were still some extant species, but this is not the case for the Palaeogene and Mesozoic, for which comparison with living organisms must be made at the level of genera. For the Palaeozoic, taxonomic affinities with the present are generally no more precise than at family level, and a large proportion of families became extinct at the end of that era.

Geological differences with the present have also increased significantly with age. Thus the relative positions of continents become progressively less certain moving back in time, as does the detailed knowledge of environmental factors such as climate and sea level. In reviewing biogeography in relation to plate tectonics, one approach has been to focus on present-day distributions of organisms and attempt to discern the historical factors which have led to such distributions (Briggs 1987).

In this book, attention is restricted almost entirely to what the fossil record can tell us, by reviewing changing distributions in relation to major geological changes, not solely associated with 'continental drift'. The major factors influencing the large-scale distribution of organisms have been outlined as far as understood, with reference to the living world but also taking into account information from the past. There follows a discussion of methods of biogeographic analysis and a brief review of some of the most significant temporal changes through the Phanerozoic, which either have a bearing on biogeography or have a biogeographic component.

The respective roles of dispersal and vicariance in controlling the distribution of organisms, have been a contentious issue in some circles. In their enthusiasm to replace the traditional interpretation involving centres of origin, associated with Darwin, some of the more zealous vicariance biogeographers have been inclined to be almost totally dismissive of models involving dispersal (see for instance Croizat *et al.* (1974) and some of the chapters in Nelson and Rosen (1981)). This seems excessive, and the more balanced view adopted here is that both vicariance and dispersal play an important role (cf. Myers and Giller 1988*b*). Thus in some cases, well documented from the fossil record, vicariance for one group of organisms has led to dispersal for others. This is clearly the case for the Pliocene emergence of the Central American isthmus linking North and South America, which led to vicariance for marine organisms and dispersal by terrestrial ones (Hallam 1981*a*).

Even an avowed vicariance biogeographer such as Brundin (1988) considers it possible in some cases to recognize centres of origin, by a marked concentration of endemic species. Some vicariance biogeographers make a distinction between *jump dispersal*, involving the crossing of barriers, and *range expansion*, involving a natural movement away from parent populations. The latter phenomenon they are prepared to accept, and would embrace the Central American case cited above, but the distinction seems a somewhat arbitrary one. Some vicariance biogeographers have been dismissive of any approach involving centres of origin, and are sceptical of the value of fossils in locating these, but who would deny today that the fossil record clearly indicates that our own species is derived from a stock that originally evolved in Africa?

The major part of this book is devoted to a narrative account of the palaeobiogeography of the Palaeozoic, Mesozoic, and Cenozoic eras. There is a case for starting with the Quaternary and working backwards through time, because our knowledge of the past diminishes significantly with increasing age. Thus it is notable that Briggs (1987) did not extend his attention back beyond the Early Mesozoic, the time when continental drift in the Wegenerian sense commenced. However, rather than running time in reverse, I have preferred the more orthodox approach of starting with the Early Palaeozoic.

While the narrative approach seems a natural one to adopt for a historical subject, the term *narrative* has been used pejoratively by certain vicariance biogeographers because of its association with scientifically untestable 'just so' stories. Such biogeographers would prefer to use cladistic methods to generate testable hypotheses for living organisms by use of the hypothetico-deductive method, and question both the phylogenetic and biogeographic use of fossils. They claim that, though fossils give evidence of minimum age, they cannot contradict the evidence of patterns obtained by living organisms. However, as Rosen (1988) points out, in an inductive framework fossils appear in a better light. Much depends on the quality of the fossil record. The further back in time, the less informative are modern distributions. Some theories of distributional change, like jump dispersal, are always available to explain awkward anomalies, because they are independent of geological processes. The problem arises of accommodating inconsistencies with *ad hoc* assumptions, producing on occasions a confusing mass of facts, ideas, and circularity, something which has bedevilled biogeography in the past.

It is possible to devise testable hypotheses using fossils, for example by comparing palaeogeographic reconstructions derived from different methods and data sets. Geology and palaeontology can interact by a process of reciprocal illumination. Where results using different groups of fossils together with geological data are concordant, confidence in interpretation is strengthened. Where they are discordant, attention can be focused on the anomalies that require further research, and on the initial assumptions that may need to be re-evaluated. This approach corresponds well with the consilience principle of one of the great pioneers of the philosophy of science, William Whewell (1840). We accept with greater confidence a hypothesis formulated to explain patterns inferred from one data set, if it also accounts for patterns inferred from completely unrelated data sets.

Ball (1983) put it well: 'historical biogeography must remain a narrative science, with pluralistic methodology and using circumstantial evidence'. In subsequent chapters it is hoped to show how a coherent and scientifically well supported biogeographic and geological history can be established by these means. Because the subject is so large it would be presumptuous to treat the following account in a book of modest size as being more than an outline.

2
Major factors influencing the distribution of organisms

The most important factors affecting the large-scale distribution of organisms are climate, the relative positions of continents and oceans, and sea level (Hallam 1981a). Before these and other factors are discussed it is desirable to pay some attention both to diversity and the varying migration potential of different types of organism.

DIVERSITY

Diversity is a measure of taxonomic variety. The two most widely used diversity indices, that combine both the number of species and their relative abundance, are the Shannon-Wiener and Simpson indices. Brown (1988) gives four reasons why their value in biogeographic studies is very limited:

1. The accurate data on population densities that are required are not usually available on a broad geographic scale.

2. The distribution of abundances within different biotas tends to be quite similar; nearly all are composed of a few common species and many rare ones. Consequently most indices give values very closely correlated with each other and with species number.

3. Rare species are usually as interesting to biogeographers as common ones.

4. The problems of differential preservation make it hazardous to apply indices to extinct biotas, and especially to compare these with extant biotas.

For these reasons species richness, a simple taxon count, is normally the measure used by biogeographers to quantify diversity. Palaeobiogeographers normally deal with generic rather than species richness, and their use of the term diversity is invariably in the sense of richness, and this is the way it is used in this book. On a spatial scale it is customary to distinguish three types of diversity: α, within communities; β, between communities; γ, large-scale, for example between different continents. Obviously it is the last category that is most relevant to biogeography.

The interpretation of diversity patterns is fraught with difficulties and remains controversial. The most favoured hypotheses of increased diversity fall into four main categories (Brown 1988): time since perturbation of environment, productivity, habitat heterogeneity, and environmental favourableness. Brown concludes that 'To account adequately for geographic patterns of diversity will require an understanding of geological history, past and present physical environments, phylogenetic history and evolutionary constraints of the taxa, the dynamics of species origination and extinction processes, and the ecological relationships of species with both their physical environment and other kinds of organisms'. A tall order, one might add, but perhaps an admittedly coarse-grained account of diversity changes over several hundred million years may at least point the way to a more satisfying interpretation.

VARYING MIGRATION POTENTIAL

The ability of terrestrial organisms to cross marine barriers varies from group to group. In a classic paper Simpson (1940) distinguished three methods by which mammals could cross from one land area to another. Free migration of a large proportion of the biota was, to a person who at that time rejected continental drift, by means of two types of land bridge, named *corridors* and *filter bridges*. Corridors allowed free migration of most organisms, whereas filter bridges 'filtered out' a significant proportion of stenotopic organisms. Thus mammals intolerant of cold were inhibited from migrating across the Bering land bridge in the Quaternary. The third method was by using a *sweepstakes route*, for example, involving chance colonization by rafting on driftwood. To obtain a winning ticket in the sweepstakes it would obviously pay to be small and breed rapidly. Rodents should be more successful than insectivores, because if hungry they could eat some of the driftwood! Simpson thought that this was the way that offshore islands as large as Madagascar, with ecologically unbalanced faunas, had been colonized. Given sufficient time, even events of low probability were likely to occur (Simpson 1952). Bats, unique among mammals in possessing the power of flight, could cross from island to island if the marine barrier were sufficiently narrow, and other mammals could possibly swim in similar circumstances. Despite their flight capacity, many continental birds will not tolerate even short marine crossings.

With regard to groups of continental vertebrates inhabiting freshwater, Myers (1938, 1953*b*) made a distinction between primary and secondary freshwater fish. The primary group consists of families rigorously restricted to freshwater, and includes lungfish. This is obviously the most useful group to use in biogeographic analysis. The secondary group includes organisms able to cross short marine barriers and reach, for example the Greater Antilles and Madagascar from adjacent continents, though not New Zealand. Amphibians are also severely restricted by seawater barriers, though frogs are comparatively mobile, one group even having reached New Zealand from Australia, according to Myers (1953*a*), though perhaps this interpretation requires re-evaluation.

It might be thought that terrestrial plants would have a greater capacity for crossing marine barriers than vertebrates. While this may be broadly true, as far back as the middle of the last century by Hooker (1859) noted that most seeds cannot be carried far by the wind and cannot tolerate prolonged exposure to salt water. It is generally recognized today that spore-bearing plants have a greater mobility than seed-bearing plants (Chaloner and Sheerin 1981).

Turning to the marine realm, it has long been evident that organisms that spend their life in the plankton tend to be more cosmopolitan than benthic organisms inhabiting the neritic zone, and the general assumption has been made that the larvae of most shallow-water invertebrate benthos cannot be dispersed over long distances (Ekman, 1953). According to Ekman, the wide, deep oceanic zone of the East Pacific is as formidable a barrier to benthos as any land barrier. Approximately 70 per cent of all benthic invertebrates possess planktonic larvae, according to Thorson (1961) who reviewed information on nearly two hundred species of temperate and boreal species and concluded that planktonic life was too short to account for larval transport across the oceans. Hence the adult species are different on the opposite sides of, for example, the North Atlantic.

However, within the tropics there are many shallow-water benthic invertebrate species that possess larvae having a planktonic stage lasting from six months to over a year, that are evidently capable of long distance transport. These have been termed *teleplanic* larvae by Scheltema (1971, 1977) from the Greek *teleplanos*, meaning far-wandering. Direct evidence from plankton samples in the tropical Atlantic shows that teleplanic larvae are found in the open sea within the upper 100 m and represent all the major 'higher' invertebrate phyla. The length of planktonic existence seems to be greater among certain tropical species than most cool temperate species, and data on current velocities shows that many larvae can be transported quite easily across the Atlantic. The larval frequency tends to decrease markedly offshore from the continental shelf, but some larvae are widely dispersed. Adults with a high frequency of larval disperal show little or no morphological difference between the East and West Atlantic.

Jablonski and Lutz (1983) distinguish among molluscs two types of larvae that can be recognized by the size of their initial shell, and are hence potentially recognizable in the fossil state. The size of this shell reflects egg size. *Planktotrophic* larvae arise from small eggs, are released in enormous numbers, and suffer huge mortality during and shortly after planktonic existence. These larvae feed on plankton and are commonly capable of prolonged free-swimming existence and thus wide dispersal. *Non-planktotrophic* larvae generally arise from large eggs, with relatively few young per parent and a generally greater parental investment and relatively low larval mortality. The larvae rely on yolk for nutrition and planktonic durations are comparatively brief, so that their dispersal capability is less than that of the planktotrophic larvae. Owing to the high dispersal capability of planktotrophic larvae, species with such larvae tend to be geographically widespread, whereas the other group is more restricted. Jablonski and Lutz also claim that, by combining larval and adult traits, it is possible to recognize modes of larval development in at least some fossil bryozoans, brachiopods, and echinoderms.

Not much information is available about the rafting transport of adult invertebrates. The most characteristically 'epiplanktonic' organisms are barnacles, hydrozoans, serpulids, and bryozoans, together with byssal, cementing, and wood-boring bivalves. Drifting material diminishes rapidly in abundance offshore and rafting is evidently restricted to relatively few species. Larval dispersal is generally much more important (Scheltema 1977).

CLIMATE

One of the most striking biogeographic features of the present day is the strong diversity gradient in most terrestrial and marine groups of organisms between the tropics and the poles (Fischer 1960; Brown 1988). Spherical harmonic analysis for numerous marine organisms indicates a systematic decline of diversity with increasing latitude, the minimum values being at or close to the poles (Stehli 1968). An example is given in Fig. 2.1 for marine bivalve species. This example also shows a significant longitudinal change at low latitude, from high values in the Indo–West Pacific region to low values in the Atlantic. Such a pattern is true for invertebrates in general.

It is natural to relate these latitudinal diversity gradients to fall in temperature with increasing latitude, and indeed many plants are confined to the tropics because they are intolerant of frost. Many of the changes in flora that take place with latitude are mirrored in tropical land areas of strongly varying relief, with increase in altitude corresponding to increase in latitude. Within the marine realm certain types of organisms, such as hermatypic corals, giant molluscs, mangroves, and many green algae are also confined to the tropics. It must of course be borne in mind that the pattern of diversity change is a general rule to which there are exceptions. Thus penguins and sandpipers are most diverse at very high latitudes and conifers, ichneumons, salamanders, and voles most diverse at intermediate latitudes.

Despite the widespread assumption that temperature is the prime controlling factor this is far from evident in many cases, particularly in the marine realm. Following the discovery of faunas in the deeper, colder Atlantic bottom waters of higher species diversity than in warmer, shallower, more inshore waters, Sanders (1968) was led to put forward a hypothesis that implied that the key controlling factor both for the latitudinal and onshore–offshore diversity gradients was environmental stability rather than temperature. Though the basis of Sanders' hypothesis has been criticized the notion of environmental stability is a useful one. High diversities within the tropics could be a consequence of the tighter niche packing of stenotopic, K-selected species, and low diversities in high latitudes may be characterized by eurytopic, r-selected species including trophic generalists. The key factor could possibly be the stability of trophic resources, these varying with season much more in high latitudes than in the tropics (Valentine 1973).

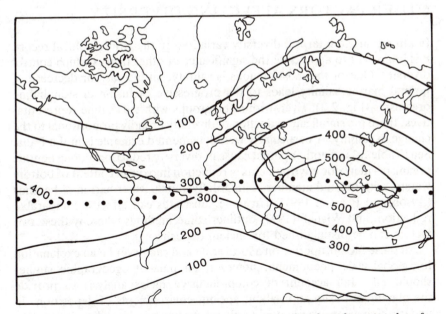

Fig. 2.1 Second order spherical harmonic surface fitted to diversity data for species of Recent bivalves. Row of dots indicates position of equator suggested by this surface. After Stehli *et al.* (1967).

Another significant factor is likely to be the incidence of light, which diminishes away from the tropics, and the greater temperature seasonality with higher latitude. Thus Ziegler *et al.* (1984) have drawn attention to the fact that most shallow shelf limestones in the Mesozoic and Cenozoic occur within the subtropics, as determined from palaeomagnetic determinations of latitude; 99 per cent of occurrences are within 45° of the palaeoequator. The limestone belt evidently did not expand polewards during the warmer climates of the Mesozoic and Early Cenozoic, suggesting that temperature is not the limiting factor. Year-round light refraction falls markedly at about 35° from the equator, the present poleward limit of Bahamian-type environments. This depositional system relies on algal fixation of $CaCO_3$, either directly or indirectly. Therefore light penetration appears to be the latitude-limiting factor influencing carbonate distribution, and there could be a corresponding influence on organisms.

Whereas thick carbonate deposits, especially those including reef limestones, are a good indicator of low latitude, warm water environments, tillites and associated glacial deposits are the best indicators of cold climates. Substantial deposits of evaporites signify not only a persistent excess of evaporation over precipitation but formation in relatively low, largely subtropical latitudes, corresponding to the present-day trade wind belts.

OTHER FACTORS AFFECTING DIVERSITY

In attempting to interpret diversity variations in the stratigraphical record it is important to appreciate the significance of other factors which control diversity. One of the most obvious is salinity; an increase or decrease of normal marine salinity leads to the progressive exclusion of stenohaline organisms (Fig. 2.2). Even within the groups with euryhaline representatives, there is a significant reduction of diversity broadly proportional to the change in salinity. A particularly well-documented example is in the Baltic Sea (Table 2.1). Another potent cause of diversity reduction among benthic organisms in neritic environments is reduction in oxygen content of bottom waters, with the end member being totally anoxic water barren of benthos (Tyson and Pearson 1991). Unstable substrates can effect a comparable reduction, and Wignall (1993) outlines some methods whereby these two factors can be distinguished for ancient deposits.

All these factors operate on a local scale and can never be an explanation for global-scale phenomena; however, all palaeobiogeographic studies should take full account of comprehensive facies analysis to provide the maximum information about ancient environments of deposition. In undertaking extensive regional or global studies corresponding facies must be compared, and the maximum diversity for a given habitat in a given region established.

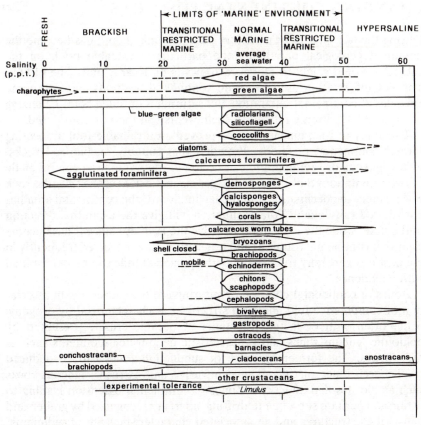

Fig. 2.2 Modern distribution of major fossilizable invertebrate and algal groups in relation to salinity. After Heckel (1972).

Table 2.1 Correlation of species diversity and salinity reduction in the Baltic Sea (based on data from Segerstråle 1957)

	Kattegat	Belt Sea	Arkona Sea	Gulf of Finland
Salinity 0/oo	35	25–30	10–15	5
Bivalves	92	34	24	4
Polychaetes	193	143	15	3
Crustaceans	232	97	29	11
Fish	75	55	30	22
Total	592	329	98	40

CHANGING CONTINENTAL POSITIONS

During the last quarter of a century plate tectonic theory has become the dominant geological paradigm; one particularly valuable book on the subject devoted solely to global tectonics is by Kearey and Vine (1990). The key information used to locate continental positions and determine episodes of continental separation and suturing is outlined here. Changing biogeographic patterns related to such events will then be considered.

The study of rock magnetism has proved an invaluable tool in locating former continental positions. Provided that suitably magnetized rocks, either igneous or sedimentary, can be found, determination can be made of two orientations that relate to the geomagnetic field at the time the rock was formed, the declination and the inclination. If the continental area has not rotated subsequently the declination will give the azimuthal direction of the pole, while the inclination will give the latitude. Since it has become clear that continental rotations and translations have occurred frequently in the past it is generally true that only the palaeolatitude can be determined with confidence.

Zones of continental collision or suturing are recognized by an association of compressive fold and fault structures as characterized by orogenic belts, often with obducted fragments of oceanic crust represented by ophiolitic igneous rocks and characteristic deep-water sediments such as radiolarites. The former presence of subduction zones can be deduced by the occurrence of calc-alkaline intrusive and extrusive igneous rocks such as diorites and andesites. Zones of continental extension leading to rifting and perhaps subsequent 'drifting' apart are recognized by graben and half-graben structures and an associated characteristic suite of sediments; any igneous rocks are likely to be dominantly alkaline basalts. New ocean floor can be recognized by the 'magnetic stripes' parallel to mid-oceanic ridges, from which the rate of sea-floor spreading can be determined if the stripes can be dated from magnetic stratigraphy.

It has become apparent that the early plate tectonic models portraying simple convergence and divergence along more or less linear or curvilinear zones were oversimplified in many cases. There has been increasing recognition of the importance of strike-slip fault movements, with sectors of continental and oceanic lithosphere moving parallel or oblique to the subduction or rifting zones, as well as normal to them. As is indicated clearly in the circum-Pacific and East Asian region, slivers of continent have been progressively accreted to major continental masses by a process sometimes termed collage or terrane tectonics. Displaced or allochthonous terranes are defined as fault-bounded geological entities of regional extent that are characterized by a geological history different from that of neighbouring terranes. They may include a complicated array of lithofacies reflecting one or more depositional environments, for example continental, oceanic, and/or island arc basins (Schermer *et al.* 1984). Palaeobiogeographic data play a vital role in the recognition of displaced terranes and in reconstructing the former positions of continents. This should become apparent in later chapters.

In evaluating the organic response to shifting continental positions and associated suturing and rifting events a fundamental principle from evolutionary biology must be adopted, that genetic isolation leads to phenetic divergence, and *vice versa*. Speciation is taken to be allopatric either in all or at least the overwhelming majority of cases. It is evident that terrestrial organisms will respond to isolation of continents by evolving independently, but the same should also be true of marine neritic organisms, to varying extents depending on migration potential and the degree of separation of the continents.

A number of simple models can be proposed that indicate the changing patterns that should develop through time (Hallam 1974). Fig. 2.3 illustrates the phenomenon of *convergence* (not to be confused with the term used in evolutionary biology). As two continents approach one another, their

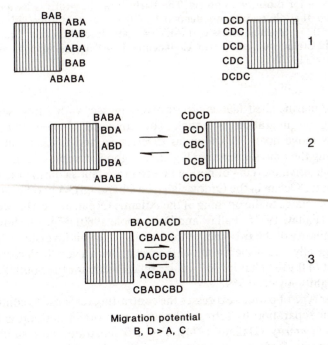

Migration potential
B, D > A, C

Fig. 2.3 Diagrammatic explanation of convergence of faunas of different continents, signified by vertical lines. As the continents approach each other as a result of sea-floor spreading, the faunas (symbolized by A, B, C, and D) progressively merge. After Hallam (1974).

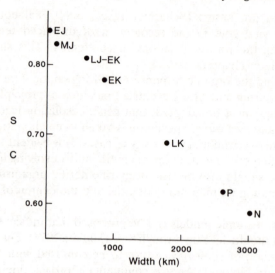

Fig. 2.4 Simpson coefficient (SC) values of faunal similarity plotted against average width of the North Atlantic Ocean basin. The LJ–EK similarity value is an average for those two epochs. The Early Jurassic width is based on the assumption that there was no deep ocean basin at that time. EJ = Early Jurassic, MJ = Middle Jurassic, LJ–EK = Late Jurassic–Early Cretaceous, EK = Early Cretaceous, LK = Late Cretaceous, P = Palaeogene, N = Neogene. After Fallaw (1979*b*).

associated marine shelf faunas progressively merge, with a time sequence depending on migration potential. The converse phenomenon, when continents move apart, is known as *divergence*, and is easily illustrated by reversing the time sequence of Fig. 2.3. A good example of convergence is the progressive merging of neritic invertebrate faunas during the closing of the Iapetus Ocean in the Ordovician (see Chapter 5). A good example of divergence relates to the opening of the Atlantic Ocean in the Mesozoic and Cenozoic (Fallaw 1979*b*; Fallaw and Dromgoole 1980). Fig. 2.4 shows how faunal similarity of the two sides of the North Atlantic is inversely related to the geologically established width of the ocean, with a very high correlation coefficient of 0.998. That for the similar relationship for the South Atlantic is only slightly lower, 0.959.

Another type of pattern expresses the contrasting response to continental suturing or separation by terrestrial and marine organisms and is known as *complementarity* (Hallam 1974). Fig. 2.5 illustrates how suturing of two continents leads to convergence of the terrestrial but divergence of the marine faunas. Good examples are the establishment of land corridors between North and South America in the Pliocene and between Africa–Arabia and Eurasia in the Miocene (Chapter 10).

A fourth pattern is termed *disjunct endemism*. This refers to a type of regionally restricted distribution of a fossil taxon in which two or more component parts are separated by a major physical barrier and hence are not readily explicable in terms of present-day geography. The classic example is *Mesosaurus*, a small reptile of Permian age that has been found only in South Africa and southern Brazil. Although possibly an aquatic form there is no doubt that it was incapable of crossing the South Atlantic. Indeed, if it were so mobile it should have spread far more widely around the world. Wegener (1924) considered this to be one of the strongest palaeobiogeographic arguments in favour of continental drift.

McKenna (1973) makes an important distinction in palaeobiogeography between what he terms Noah's Arks and Beached Viking Funeral Ships. Noah's Arks are continental fragments that transport living biota, with resulting convergence of terrestrial faunas following collision with another continent. Viking Funeral Ships transport long-dead organisms in the form of fossils from one continent to another, with no biological consequences but potential for great confusion in palaeogeographic reconstructions. This

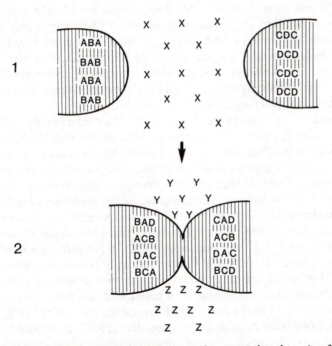

Fig. 2.5 Complementarity in the distribution of terrestrial and marine faunas as two continents approach each other and eventually become sutured together. The terrestrial faunas exhibit convergence because of the creation of a corridor, while the marine organisms diverge from common ancestors X to faunas Y and Z. After Hallam (1974).

notion is of great value in helping to understand the complications of collage tectonics, involving as it does the investigation of allochthonous terranes.

SEA-LEVEL CHANGES

The stratigraphic record gives abundant evidence of rises and falls of sea level with respect to the continents throughout the Phanerozoic. Some of these were global in extent, others merely regional, but in either case the biogeographic consequences for both marine and terrestrial organisms must have been significant (Hallam 1992). At present, climate is a major control on biogeographic distributions because of the marked temperature gradient from the tropics to the poles, but in more equable times in Phanerozoic history the role of climate is likely to have assumed lesser importance. Expansions and contractions of epicontinental seas would have been correspondingly more important in controlling the degree of endemism of marine faunas. At times of low sea level and restriction of seas, faunal migrations between continental shelves would be rendered more difficult, and with less gene flow more local speciation would take place among the less dispersible organisms that occupied shallower water habitats. Conversely, at times of high sea level there should be a reduction in marine endemism and greater spread of cosmopolitan species. Rise of sea level does not always correlate with reduction of endemism, though it appears from the stratigraphic record to be the general case; much depends on the regional palaeogeographic situation.

Changes of sea level can also have an important effect on the distribution of terrestrial organisms, even without lateral movements of continents. It is a matter of indifference to them whether the sea separating them from their relatives on another landmass is underlain by oceanic or continental crust. Thus at times of high sea level emergent regions of continents may be isolated by epicontinental sea, leading to the evolution of independent terrestrial biota. Good examples are the North American mid-continent sea in the Late Cretaceous and the Eurasian Turgai Sea in the Palaeogene.

At the present day, and presumably also in the past, the richest biota in terms of both diversity and biomass occur in the neritic zone down to about 200 m, and some that depend on light, such as algae and hermatypic corals, are confined to the shallower reaches of this zone. Nevertheless a high proportion of phyla also have deep-sea representatives (Fig. 2.6).

It was a surprise when Hessler and Sanders (1967) discovered that in the North Atlantic the species diversity of polychaetes and bivalves was greater on the continental slope than on the shelf, with a gradient of increasing diversity from inshore waters. This result led directly to Sanders (1968) putting forward his stability–time hypothesis. The diversity gradient has, however, been reinterpreted by Abele and Walters (1979) as being merely a

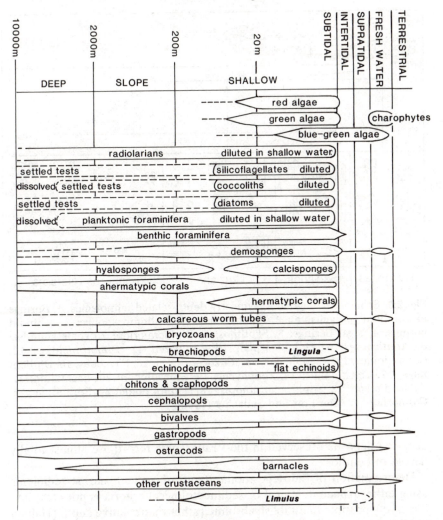

Fig. 2.6 Modern distribution of major fossilizable invertebrate and algal groups in relation to depth of sea. After Heckel (1972).

consequence of the well-known species–area effect, with the size of sample area increasing progressively towards the deep sea.

Detailed palaeoecological research on benthic invertebrates indicates that throughout the Phanerozoic there has been an increase in species diversity from inshore, environmentally stressed communities to offshore communities in more stable environments (Bambach 1977). All these studies concern the neritic zone, where the vast majority of benthic invertebrates have lived, and the relatively few examples of fossiliferous

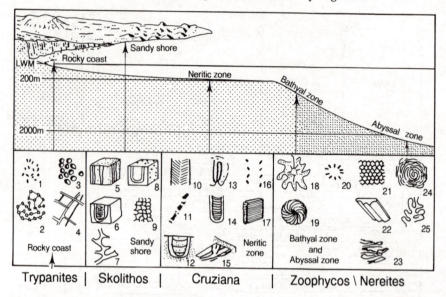

Trypanites | Skolithos | Cruziana | Zoophycos \ Nereites

Fig. 2.7 Schematic diagram of typically depth-related ichnofacies. 1, Borings of *Polydora* (polychaete); 2, *Entobia*, boring of clionid sponge; 3, echinoid borings; 4, algal borings; 5, *Skolithos*; 6, *Diplocraterion*; 7, *Thalassinoides*; 8, *Arenicolites*; 9, *Ophiomorpha*; 10, *Cruziana*; 11, *Dimorphichnus*; 12, *Corophioides*; 13, *Rusophycus*; 14, *Rhizocorallium*; 15, *Phycodes*; 16, *Diplichnites*; 17, *Teichichnus*; 18, *Zoophycos*; 19, *Spirophyton*; 20, *Lorenzinia*; 21, *Palaeodictyon*; 22, *Taphrhelminthopis*; 23, *Helminthoida*; 24, *Spiroraphe*; 25, *Cosmorhaphe*. After Frey and Seilacher (1980).

deep-sea deposits preserved in the stratigraphic record are almost totally devoid of benthos.

Determination of the depositional depth of ancient marine sediments using either palaeontological or sedimentological criteria is not easy, and it is rarely possible to establish absolute rather than relative depth (Hallam 1967*b*). The main problem is that depth of water *per se* is not an environmental factor and the best one can hope to do is to seek a relationship with some depth-related factor such as temperature, pressure, water energy, or incidence of light, which does have an environmental influence. Such a relationship is not necessarily simple. Thus palaeontologists have used the ratio of planktonic to benthic foraminifera in the Late Cretaceous and Cenozoic as a measure of relative depth. This is because benthic foraminifera are more abundant today in inshore shallower water of the continental shelves and planktonic foraminifera more abundant in offshore, deeper water. However, around oceanic islands the planktonic: benthic ratio may be high, indicating that the key factor is not depth but pelagic influence.

Trace fossils were thought to be one of the best types of bathymetric indicator and provided the basis for the well-known depth zonation of Seilacher (1967 and Fig. 2.7). However, the type ichnogenus of the *Zoophycos* facies, thought to signify bathyal depths intermediate between the neritic and abyssal zones, is a common element in Palaeozoic, though not younger, shallow neritic faunas. Furthermore, vertical burrows characteristic of the shallow neritic *Skolithos* facies can be found, albeit uncommonly, in the present deep sea, and the complex sediment surface-parallel burrows and trails of the 'abyssal' *Nereites* facies can be found in strata whose other characteristics indicate deposition in relatively shallow, though quiet, water. This example should illustrate that bathymetric determinations are not a straightforward matter, but generally in palaeobiogeographic research it is possible to achieve first-order approximations that are adequate for the purpose.

With regard to organisms that lived on the continents problems are much greater. The fossil record is more patchy because only zones of low topography such as lakes and river floodplains have a depositional record and determinations of palaeoaltitude in mountain or plateau zones are usually mere guesswork. This can pose serious difficulties in interpreting, for instance, the former living environments of plant fossils, because many plant remains are dispersed easily by wind or water and could have come from uplands. Similarly vertebrate fossils have often been transported into and even along river channels, one of the best sites for preservation. Once again, it should be remembered that palaeobiogeographic studies normally concern themselves with gross distributions over extensive regions, unlike palaeoecology or taphonomy.

3
Methods of biogeographic analysis

It would be inappropriate in a book of this sort to deal with analytical methods in any great detail, but it is desirable to address a number of relevant matters before turning to the temporal and spatial distributions of organisms.

BIOGEOGRAPHIC SUBDIVISIONS AND THE SPECIES–AREA RELATIONSHIP

A variety of terms have been used for global biogeographic subdivisions. The nineteenth century pioneer biogeographers favoured *region*, to which have been added subsequently *realms* and *provinces*. Schmidt (1954) recognized four categories in a hierarchy: realm, region, subregion, province. The term region has fallen into disuse, and current biogeographers are content to use realm, for larger, and province, for smaller categories. If further subdivisions are required then terms such as subrealm or subprovince can be introduced, but this is not widespread practice. Such terms should be purely descriptive and carry no environmental connotations. Thus the proposal by McKerrow and Cocks (1986) to distinguish provinces, separated by barriers, from climatically controlled realms has little to recommend it.

Sometimes attempts have been made to use ecological categories, such as *biome*, in a biogeographic classification (Sylvester-Bradley 1971). For example in the marine realm one can distinguish coral reef, level-bottom benthic and planktonic biomes, or many others, but this type of classification seems irrelevant to biogeography, which is concerned with the distribution across the globe of similar types of biome, or the geographic restriction of some biomes. Thus meaningful questions can be asked about why the coral-reef biome is restricted to the tropics, or about why the diversity in level-bottom biomes diminishes with increasing latitude.

Sporadic attempts have been made to quantify biogeographic classifications. Kauffman and Scott (1976) defined a province as containing between 25 and 50 per cent of endemic species. There are a number of difficulties about such an approach, of which the most important one concerns the distribution of species occurrence. Many sets of data about the numbers of individuals in each species fit logseries or lognormal distributions, and the same distributions have been shown to apply to the distribution of species occurrence (Buzas *et al.* 1982). Most species are not abundant, and rare species should not be used in biogeographic studies, because their absence could be misleading. Considering Kauffman and Scott's definition, the amount of endemism required to define a province could be reached simply through the contribution of rare species, which could give a misleading picture of the biota as a whole.

There is, in fact, no consensus on how to define a province or realm in a way which would be valid for widely different organisms and times in geological history. One is obliged to depend on the consensus of specialists working on particular groups and periods, and even in this situation it is remarkable how much can be communicated successfully in purely qualitative terms. This is not to deny the desirability of quantitative treatment of biogeographic data in appropriate circumstances.

One of the most important biogeographic generalizations that has emerged from numerous studies of living organisms is the so-called species – area relationship (Connor and McCoy 1979; Williamson 1988). This relationship is an exponential one as given by the following equation:

$$S = c\,A^z$$

where S is the number of species, A the area, and c and z fitted constants.

That the relationship probably applies to higher taxonomic levels is suggested by the study of Flessa (1975) who showed that mammalian generic, familial, and ordinal diversity correlate significantly with continental area, suggesting that the area effect is similar in form to that shown for true islands. What might be described as the taxon–area relationship is likely to have been valid for the past, and indeed Flessa attempts to show its relevance to diversity changes among Late Cenozoic mammals following the emergence of the Panama land bridge linking North and South America.

There is no general agreement about the explanation of the taxon – area effect, though the most plausible one implicates environmental heterogeneity in larger areas (Williamson 1988). McArthur and Wilson's (1967) theory of island biogeography has been of great influence, and has been heuristically valuable in experimental work (Schoener 1988). The main premise of the theory is that each island is in a state of dynamic equilibrium between species immigration and extinction. This is in contrast to the traditional view that the low species diversity of small islands signifies a depauperate or 'impoverished' condition, implying an unfavourable environment. Despite the great stimulation to research that this theory has produced its validity is controversial. Not all experimental studies support the quantitatively predicted relationships (Schoener 1988) and Williamson (1988) reckons that it offers no helpful answers to some of the major questions concerning the species–area relationship, such as the slope of the curve.

QUANTITATIVE TECHNIQUES

The most widely used quantitative techniques for comparing biota of different regions are the calculation of binary (presence–absence) similarity coefficients, namely the Jaccard, Dice, and Simpson (Cheetham and Hazel 1969).

The Jaccard Coefficient is $(C/N_1 + N_2 - C)$ 100, where C is the number of taxa in common and N_1 the sample with the smaller number of taxa. The Dice Coefficient is $(2C/N_1 + N_2)$ 100 and the Simpson Coefficient (C/N_1) 100. The Simpson Coefficient, which has proved the most popular among palaeobiogeographers, was devised to minimize the effect of unequal size of the two biota being compared: the denominator has only the number of taxa in the smaller sample.

If the two faunas are of equal size, the Jaccard and Dice Coefficients are suitable, but if there is a big discrepancy the larger fauna would distort the resulting value so that the degree of relationship would be obscured. On the other hand the Simpson coefficient may give unstable results if one sample is very small, because small increases of information may produce disproportionately large effects on the result (C.P. Hughes 1973). Nevertheless, in his study of trans-Atlantic Mesozoic and Cenozoic invertebrates Fallaw (1979a) found that the results derived from Simpson Coefficient calculations correlated well with observations by various authorities and with geological predictions of a widening ocean based on sea-floor spreading chronology. Values calculated from the other two binary coefficients either contradicted or were only weakly supported by other evidence.

Shortcomings of all these indices have been pointed out by a number of researchers. They have not been derived in a mathematically rigorous way and their validity has all too often been tested by whether or not they seem to work in practice. Furthermore, they have not been tied to clearly defined null hypotheses. In consequence statistically meaningful comparisons between values of a coefficient are impossible. Henderson and Heron (1977) and Raup and Crick (1979) have attempted to resolve this problem by devising rigorous and statistically valid similarity measures.

One other binary method applied in palaeobiogeography should be mentioned here, namely Johnson's (1971) Provinciality Index PI, where:

$$PI = C/2E$$

with C being the number of genera in common between two regions and E being the number of genera endemic to the region of lower diversity. This provides a straightforward method for comparing regions with samples of differing sizes. PI = 1.0 where 50 per cent of all genera are cosmopolitan if the samples are of the same size (or normalized to twice the smaller fauna if the sample sizes differ). PI = 0.5 where the number of cosmopolitan genera equals the number of endemic genera in the smaller fauna. Though the index is admittedly not ideal, it does provide a method of comparing samples of differing size by normalizing to the smaller sample.

In addition to the above methods for determining biotic distance there are others that endeavour to simplify the original data by seeking groups of samples that are relatively similar. Single link cluster analysis is the soundest method analytically and the one least likely to give misleading results (C.P. Hughes 1973). Clusters can be evaluated further by discriminant analysis.

There are also methods of co-ordinate representation, which seek ways of reducing the number of attributes needed to describe the samples and hence reduce an *n*-dimensional configuration to one of a small number of dimensions, conveniently two or three for ease of pictorial representation and comprehension. Q-mode principal components analysis has become popular in palaeobiogeography because it generates simple principal components of one or a few taxa that normally dominate samples and has proved useful in mapping geographic distributions, of for example Quaternary planktonic foraminifera (Imbrie and Kipp 1971). Of the various methods tested by Malmgren and Haq (1982) using Atlantic coccoliths, the most efficient were found to be maximum-likelihood factor analysis with varimax rotation, and R-mode principal components analysis, also with varimax rotation.

Because in palaeontology there are many data based on inadequate collections and doubtful taxonomy, C.P. Hughes (1973) believes that the analytical method utilized should be as robust as possible, because it will then be least affected by inadequacies of the data. He favours a method called non-metric multidimensional scaling. An example of its use on Ordovician trilobite distributions is provided by Whittington and Hughes (1972).

This section concludes with a reference to the possibility of sampling bias. Sample size effects are potentially large because most species occur very infrequently. Koch (1987) undertook an analysis of ten large data sets, each based on over a thousand occurrences, of Holocene and Cretaceous foraminifera and molluscs. All ten data sets fit Fisher's log-series distribution well, and the effect of sample size can now be predicted for a wide range of studies. If sample-size effects are not explicitly considered, minimum information on the number and size characteristics of palaeontological data sets are necessary in order to allow other workers to evaluate their reliability.

In her studies of Palaeozoic plants Raymond (1985) found that the *average* generic diversity, defined as the number of genera in each assemblage divided by the total number of assemblages, is less sensitive to sampling bias than *total* diversity, defined as the number of genera present in each region, or globally during each time interval. This is because poorly sampled regions and time intervals generally appear to have lower diversity than well sampled ones. Such biases have their greatest effect on small samples.

Belasky (1992) undertook a test of the probabilistic method of palaeobiogeographic analysis of Henderson and Heron (1977), which provides estimates of population diversity at a site and attempts to determine actual distributions of taxonomic diversity on the basis of common palaeobiogeographic data. He utilized data on the occurrence and diversity of 90 living hermatypic coral genera in 165 areas in the Indian and Pacific oceans. The method was shown to be successful in reducing sample bias. A sample represents, on average, 84 per cent of the population. The sampling bias in most palaeobiogeographic data sets is expected to be at least as high, that is 16 per cent.

Few palaeobiogeographic studies have undertaken data analysis with the maximum quantitative rigour possible. This does not necessarily invalidate their conclusions because much depends on the adequacy of the data to resolve particular questions, but it does indicate the opportunity for improvement.

METHODS INVOLVING VICARIANCE BIOGEOGRAPHY

Panbiogeography is a biogeographic method introduced by Croizat (1952, 1958, 1964) that focuses on the spatiotemporal analysis of distribution patterns of organisms and is distinct from phenetic biogeography, which investigates similarities between biotas in terms of the number of taxa in common (Craw 1988). Croizat's method involved drawing lines on a map, known as tracks, which connect the known distributions of related taxa in different areas. When two or more tracks of unrelated taxa coincided, he combined them into *generalized tracks*. According to Croizat, a generalized track indicates the distributional pattern of an ancestral biota before it vicariated, because he was able to show that biotic distributional patterns were non-random and fell into a limited number of tracks. The tracks for terrestrial organisms may cross oceans and hence could not be explained by present-day geography.

With the general acceptance of plate tectonics, and hence continental drift, by the early 1970s it was quite natural that Croizat's voluminous but obscure publications were seized upon and used to provide a foundation for a new school of vicariance biogeography (Croizat *et al.* 1974). This paper, and subsequent ones involving the co-authors Nelson and Rosen, were characterized by a polemical approach that virtually denied any validity to the alternative, dispersalist school which had hitherto held sway. Many have subsequently reacted against this excessive polarization, and dismissive attitude towards dispersalist mechanisms, but without question a new scientific rigour has been introduced, with more emphasis being placed on testing models and with *ad hoc* hypothesizing being discouraged.

An important difference quickly emerged between Croizat and his co-authors and other vicariance biogeographers, who also supported cladistic methods of taxonomic analysis, whereas Croizat utilized a conventional phenetic method. Brundin (1988) favours a balanced approach utilizing cladistic analysis to establish phylogenetic relationships but accepting an important component of dispersal. Indeed, the achievement of cosmopolitanism by dispersal, whether across barriers or by range expansion, is a fundamental prerequisite for the widespread occurrence of harmoniously diversified biotas that may subsequently be subjected to vicariance events. Using chironomid midges to illustrate his point, Brundin showed how, by considering relative apomorphies in conjunction with known geological history, one can in many cases determine paths of dispersal and pinpoint areas of origin of particular taxa. Phylogenetically primitive members of a group will by definition be found near the centre of origin, and 'derived' or youngest members on the geographic periphery. This is in effect what is stated in the so-called progression rule of Hennig, the founder of cladistics; the most apomorphic taxa are those furthest away from the centre of origin, which contains the most plesiomorphic taxa.

It seems evident from recent research that both cladistic taxonomy and vicariance biogeography are here to stay, and an influential attempt to combine them has been made in the form of *cladistic biogeography* (Nelson and Platnick 1981; Humphries and Parenti 1986). The method first introduced by Platnick and Nelson (1978) involves the initial finding of monophyletic groups with taxa occurring in three or more similar areas. Cladograms are produced for the taxa and the names of taxa at the terminal tips of the cladograms replaced by the names of areas in which each taxon occurs. These are called areas cladograms (Fig. 3.1). The sum of the areas on one cladogram is equivalent to a track, in Croizat's terminology. The corroboration of a particular pattern in a consensus cladogram or general area cladogram is the best hypothesis that can be obtained to express the relative recency of biotas. Each cladogram is never quite the same as another as a result of 'missing', 'redundant', 'unique', and partly overlapping areas. Two general approaches to solving this problem are outlined by Humphries *et al.* (1988).

Cladistic vicariance biogeographers have confined their attention to living organisms and the use of fossils in both phylogenetic systematics and historical biogeography has been challenged. However, as Grande (1985) points out, fossils assume primary importance in four different ways:

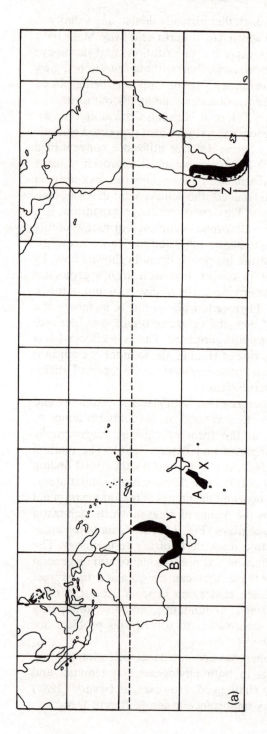

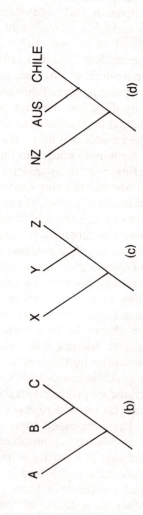

Fig. 3.1 Hypothetical distribution of two groups—fishes (A, B, C) and flowering trees (X, Y, Z). (b, c) Cladograms for each group. (d) Area cladogram common to both groups. After Humphries and Parenti (1986).

(1) they provide additional taxa to give new morphological and ontogenetic data;

(2) they provide additional taxa to increase the known biogeographic range of a taxon;

(3) they help to establish a minimum age for a taxon;

(4) they present ancient biotas which can be examined for biogeographic patterns not recognizable in younger or older biotas.

The latter two ways are unique to fossils.

Nelson and Platnick's method does not address the problem of areas with complex biotic histories, whose biogeographic affinities change with time. If we accept the possibility that some species disperse, then we must also accept that in areas where sufficient dispersal has occurred, it may be difficult or impossible to recognize an older area of endemism from Recent biota. The Recent biota of certain areas could be the biota least likely to show a fully resolved biogeographic pattern because it may contain certain descendents of older endemic components together with species that later dispersed into the area or were introduced by humans. The later dispersals, with extinctions, could eventually mask the pattern of the older endemic components.

Grande proposes a time-controlled vicariance method and gives an example utilizing freshwater teleost fish, which have a good fossil record, being more numerous, better preserved and more complete than most other vertebrates in an aquatic environment. The teleosts of the Eocene Green River Formation in western North America show a biogeographic pattern not visible in younger biotas, and evidently show Australasian affinities. Fig. 3.2 shows area cladograms based on biological groups and also on geological evidence, to illustrate how his method may be applied.

However, as a general rule, there are serious drawbacks about applying the cladistic vicariance method to palaeobiogeography. In the first place, it requires good cladistics leading to the clear recognition of monophyletic groups, and this is not available for most fossils at present and perhaps for the foreseeable future as well. Furthermore the method can only generate geological hypotheses of area and biotic divergence. Although practitioners have challenged this criticism the alternative of convergence is generally inferred indirectly (Rosen 1988). It should also be recognized that area cladograms are not true cladograms because they are not derived from the explicit analysis of geographic or geological characters in a way analogous to the way synapomorphies are analysed in systematics.

These problems can be overcome by use of a method called *parsimony analysis of endemicity* (PAE) (Rosen and Smith 1988). This method takes as its starting point presence–absence data for a set of sample localities and particular organisms. Shared presence (endemicities) are

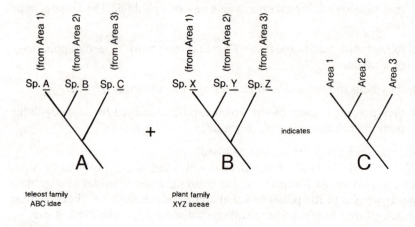

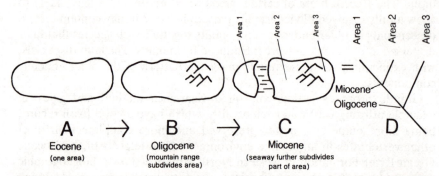

Fig. 3.2 Upper diagrams, area cladogram (C) based on biological evidence (A and B). Lower diagrams, area cladogram (D) based on geological evidence (A, B, and C). After Grande (1985).

treated analogously to synapomorphies in systematics, with identification of nested hierarchical patterns as in cladistics. Corroboration of hypotheses sought in analogies of many groups has the advantage that it can be applied to groups whose phylogenetic relationships are unspecified. Recognition of congruent patterns in different groups of organisms is strong evidence of the same historical events. Fig. 3.3 illustrates how parsimony analysis of endemicity treats a mixed history of convergences and divergences. PAE cladograms always produce a nested series of biotas, based only on how recently they have had taxa in common, irrespective of the nature of the historical events. Rosen and Smith point out that it is fallacious to use cladograms as a direct record of vicariance events. The final step is to translate the results into geological history, termed TECO events by Rosen (1984)—tectonic, eustatic, climatic, and oceanographic events.

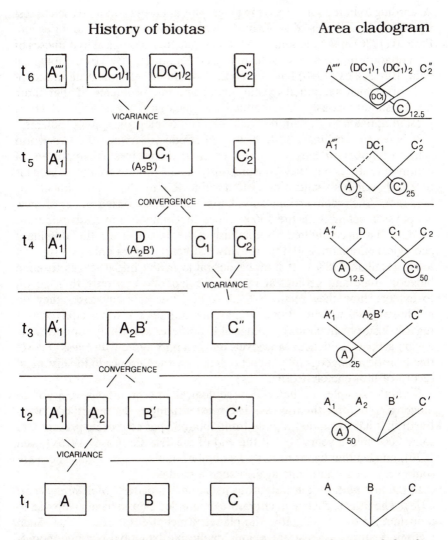

Fig. 3.3 A hypothetical history of biotas and the corresponding sample locality cladograms. t_1–t_6 = successive time planes. Geographic sample localities are depicted as rectangles, biotas as letters. The initial faunas A, B, C have no taxa in common. Other symbols : C, C^1, C^{11}, etc, turnover of biota C through time; D, endemic element of biota developing after convergence event producing A_2 and B^1 faunal mixing. In the area cladograms the common elements of biota shared between sister localities are circled, and the percentage of the total biota shared at the vicariance event that has survived to this time, is also given. After Rosen and Smith (1988).

BIOLOGY VERSUS GEOLOGY

According to Patterson (1983) biogeographic analysis cannot be used to test rigorously hypotheses of geological events and *vice versa*. Humphries and Parenti (1986) make the same point and argue that geological hypotheses of area relationships are as reliable as area cladograms derived from biological data. But as Cox (1990) has observed, it would be self-defeating to accept a theory which simplified existing biogeographic problems, if that same theory simultaneously disassembled the geological framework which at present appears to explain the vast majority of biogeographic problems.

It is worth recalling a little history of science. Between the two world wars, a number of biogeographers counted themselves as supporters of continental drift, but they conspicuously failed to convert the consensus of Earth scientists and were often subjected to the same ridicule as Wegener. Such conversion only happened as a consequence of new geophysical research in the 1960s. Since that time an impressive amount of data has accumulated to establish plate tectonics as the dominant geological paradigm, and it thoroughly justifies its almost universal support today (Hallam 1983*a*). It may be desirable in the initial stages to keep biogeographic and geological hypotheses separate and only then go on to explore how they relate to each other, but in many cases they do not have equal weight. Biogeographic hypotheses are often equivocal as regards history while many geological hypotheses are solidly supported by a wide array of evidence. Moreover, one can only agree with Flessa (1981) that historical biogeography is best practised with reference to the historical record of life, namely fossils.

Several examples of heterodox geological interpretations relevant to biogeography offer themselves. The most striking is the hypothesis of an Earth that has expanded rapidly from about 80 per cent of its present size some 200 million years ago, at the end of the Triassic (Carey 1976; Owen 1976), an idea that has appealed to a number of biogeographers. There are some very solid arguments against such a model.

In the first place it fails a rigorous palaeomagnetic test (McElhinny *et al.* 1978). The drastic change in curvature demanded should have given rise to a distinct set of cracks across the planet, which are not observable. Such tectonic features as exist are readily explicable by global tectonic models that require an unchanging Earth diameter. The expanding Earth model denies the reality of large-scale subduction, for which there is abundant evidence. It also demands a drastic fall of sea level since the Triassic, because the expansion must have been expressed in the oceans rather than on the continents. In fact sea level rose significantly to a Late Cretaceous peak (Hallam 1984*a*). These are also geophysical problems which expansion would provide in requiring rapid reduction of the rate of rotation.

Owen's (1976) attempted reassembly of the continents has attracted considerable attention. He found that such a reassembly into the super-continent Pangaea resulted in a series of V-shaped gaps between the continents, which only disappear if the Earth has expanded by the amount specified. Thus there is no Arctic basin or Tethys embayment, and never any great distance between Australia and south-east Asia. However, Owen's methodology has been shown to be erroneous by Weijermars (1986). If a three-dimension globe is used rather than map projections, no such gaps appear.

So what are the arguments that persuade some biogeographers to favour Earth expansion? Estes (1983) welcomed Owen's reconstruction to explain the strong similarities in the lizard populations of Australia and south-east Asia. But, as Cox (1990) notes, recent geological work shows that the gap between these two regions was much smaller than shown in earlier plate-tectonic tectonic reconstructions.

A more extreme expanding-Earth hypothesis has the Pacific Ocean opening in the Jurassic, based on similarities among various taxa in eastern Asia and North America. The analysis by Fallaw (1983) of marine fauna's changing degree of similarity through time indicates, however, a post-Triassic trend of increasing ease of migration, in accord with a conventional plate-tectonic model of decreasing width of ocean. Similarities of continental biota can likewise be rationally accounted for without recourse to opening the Pacific (Cox 1990).

Another heterodox geological concept that has proved popular among some biogeographers is that of Pacifica, a continent adjacent to Australasia before the Late Permian, which fragmented in the Mesozoic, with the fragments being dispersed to the Pacific borders of the Americas and east Asia (Nur and Ben-Avraham 1977, 1981). However, there is little evidence that the circum-Pacific terranes are of continental origin, and no evidence of common characteristics to suggest a single primordial continent. Nearly all the constituent rocks represent old sea mounts, island arcs, or slivers of deep ocean floor (Schermer *et al.* 1984). A further difficulty pointed out by Cox (1990) is that some of the biogeographers in question concern themselves with geologically young groups such as mammals and angiosperms, which have only radiated significantly to become dominant organisms within the last one hundred million years. Thus the component fragments of the purported Pacifica could not have borne specialized representatives of these groups.

Biogeographers have long puzzled about the problem of amphitropical distributions; of taxa present in the northern and southern hemispheres, but absent from the tropics. The area cladograms of Humphries and Parenti (1986) repeatedly show at least two associations, with the closest relatives of austral groups being found in the boreal zone (for example the southern beech *Nothofagus* and the beech *Fagus*). Since it is considered

and cool modes, with the third cool mode being in the Mid-Jurassic to Early Cretaceous time interval (Fig. 4.1). This is surprising in that the Mesozoic has generally been considered a relatively warm, equable time of Earth history, and is based on the presence in high palaeolatitudes of phenomena suggesting the likely existence of polar ice, such as dropstones and glendonites. There is, however, absolutely no indication of extensive ice sheets, and plenty of evidence from fauna and flora for relatively mild conditions in high latitudes compared with the present day.

The general trend of eustatic sea level is also shown in Fig. 4.1 (Hallam

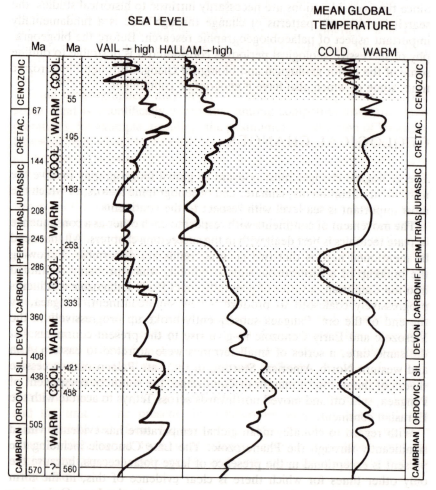

Fig. 4.1 Approximate and qualitative sea-level and temperature curves for the Phanerozoic. Stippled zones represent postulated cooler periods. Simplified from Frakes *et al.* (1992).

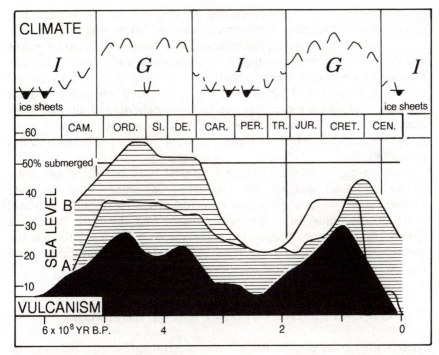

Fig. 4.2 The two supercycles of Phanerozoic history. I, icehouse state, G, greenhouse state. Sea levels (A) according to Vail *et al.* and (B) Hallam. Volcanism as judged from American granite emplacement. Simplified from Fischer (1984).

1992). Sea level rose through the Early Palaeozoic to a Mid Ordovician peak and thereafter showed an oscillatory decline to a low at the end of the Permian. There followed a progressive rise to a Late Cretaceous peak and subsequent fall to the present abnormally low level, matched only by the end-Palaeozoic low.

An attempt to link the three phenomena, plate tectonics, climate, and sea level, has been made by Fischer (1984), who has proposed a model of two Phanerozoic supercycles, with the climate alternating between 'greenhouse' and 'icehouse' states (Fig. 4.2).

Times of rapid sea-floor spreading (and corresponding subduction zone activity) should correspond to times of high sea level because of increased ocean ridge volume. High sea level reduces the area of continents and, hence, the amount of carbon dioxide taken up by weathering. In consequence, high sea level should correlate with high atmospheric CO_2 and a tendency towards global equability, low sea level with low CO_2 and the growth of polar icecaps.

For first-order eustatic changes Fischer used two previously published, broadly similar sea-level curves and for the amount of volcanism a compilation of Phanerozoic plutonic activity in North America, thought to be a good proxy. As supercontinents have been broken up by rifting and dispersed, the mean continental thickness has diminished, facilitating flooding by the ocean, whereas coalescence of smaller continents has led to increased thickness. Increased volcanism associated with active sea-floor spreading introduces more CO_2 into the hydrosphere, and ultimately the atmosphere, thereby leading through the greenhouse effect to rise in mean global temperature, which reinforces the trend referred to in the previous paragraph.

Fischer's model is attractive in a general way, but there are some awkward matters that are not adequately explained. Thus the Late Ordovician Gondwana glaciation occurred at a time of high sea-level stand, in the midst of a greenhouse state, and the Late Palaeozoic Gondwana glaciation disappeared after the Mid Permian, though the supercontinent Pangaea remained coherent until well into the Jurassic and sea level remained comparatively low.

MAJOR PATTERNS OF DIVERSITY CHANGE

After some earlier dispute there is now general agreement that diversity has increased significantly through the course of the Phanerozoic, and at an accelerating rate in the Mesozoic and Cenozoic (Sepkoski *et al.* 1981; Valentine 1985).

The marine record of animal life is by far the best documented. Raup (1976) attempted to establish species richness on the basis of citations in the *Zoological Record*. His data presented in Fig. 4.3 are dominated by organisms with a high preservation potential, from deposits of epicontinental seas and continental margins, so that, for example, insects are only about 3.5 per cent of the total. Fig. 4.3 also shows a curve of species richness increasing at a greater rate, put forward by Allison and Briggs (1993) to adjust for palaeolatitudinal bias. As is well established, species diversity decreases with increasing latitude. Almost all marine sedimentary rocks in the Palaeozoic of Europe and North America, where 54 per cent of described fossil species were collected, were deposited in the tropics, whereas only 24 per cent were deposited in such low latitudes in the Mesozoic and Cenozoic. Allison and Briggs' plot indicates a Palaeozoic acme in the Permian rather than the Devonian, and the Triassic recovery following the end-Permian mass extinction is much less pronounced than suggested by previous studies. Bambach (1977) has studied invertebrate species richness in benthic habitats and concluded that within-habitat species richness has increased by a factor of four since the Mid-Palaeozoic.

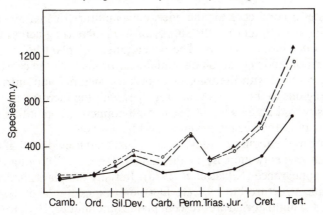

Fig. 4.3 Species richness for the Phanerozoic. Solid line : based on Raup (1976). Dashed lines; adjusted to account for palaeolatitudinal bias, assuming 33 per cent (coarse) and 10 per cent (fine) endemicity. After Allison and Briggs (1993).

A similar trend is shown for marine metazoan families (Fig. 4.4). Sepkoski's (1981) factor analysis reveals what he calls three 'evolutionary faunas' labelled respectively I, II, and III. Fauna I dominated in the Cambrian but declined from Early Ordovician times onwards. Fauna II rose to importance in the Ordovician but declined drastically after the end-Permian mass extinctions. Fauna III was a subordinate element of Palaeozoic faunas but overwhelmingly dominant in the Mesozoic and Cenozoic.

The overall pattern of the striking diversity increase in the Mesozoic and Cenozoic is attributed by Valentine (1973) mainly to the combined effects of plate tectonics and climate, the former predominant in the Mesozoic and the latter predominant in the Cenozoic. Both continental breakup and the greater differentiation of latitude-parallel climatic zones will have resulted in increased endemism and hence higher diversity. The Early Palaeozoic diversity increase is a likely consequence of the spectacular Cambro–Ordovician metazoan radiation, associated to some extent with the splitting up of a Proterozoic supercontinent. With regard to the post Mid-Palaeozoic increase in within-habitat species richness, Bambach (1977) speculates that it may relate to increase in availability of food resources, such as the developing terrestrial flora.

In a more recent study Bambach (1993) argues plausibly that the biomass of marine consumers has increased during the Phanerozoic, as indicated both by the increase in fleshiness and average size of individuals of dominant organisms, and by a conservative estimate that dominant organisms in the Cenozoic are at least as abundant as those in the Palaeozoic. Bambach also makes a good case for the energy expenditure of marine consumers increasing as well, and for the supply of food increasing across the whole spectrum of marine habitats. The development of plant life on land and the impact of land vegetation on stimulating productivity in coastal marine settings, coupled with the transfer of organic material and nutrients from coastal regions to the open ocean, together with the increase through time in diversity and abundance of oceanic phytoplankton, point to increased productivity in the oceans through the Phanerozoic.

Turning to the fossil record of vascular plants, it is evident that there was significant diversity increase in the Late Palaeozoic following their Late Silurian appearance (Niklas *et al.* 1985). Initially the plants had small size and simple structure but by Late Devonian times the beginnings can be discerned of multistoried arborescent plant communities. By Carboniferous times, marked by higher diversity than in the Devonian, large lycopsid and sphenopsid trees had become established, together with many normal ferns, seed ferns, and conifers. By the Mid Jurassic, some 80 per cent

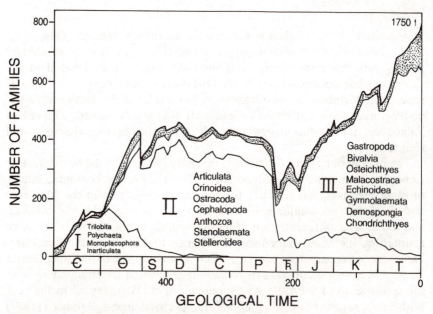

Fig. 4.4 The history of Sepkoski's three evolutionary faunas as components of total marine diversity. The uppermost curve shows the total number of metazoan families known from the marine fossil record through Phanerozoic time. The stippled field immediately below the curve for total diversity represents the residual diversity not accommodated by the three factors that allow discrimination of the evolutionary faunas. C, Cambrian; O, Ordovician; S, Silurian; D, Devonian; C, Carboniferous; P, Permian; TR, Triassic; J, Jurassic; K, Cretaceous; T, Tertiary. After Sepkoski (1981).

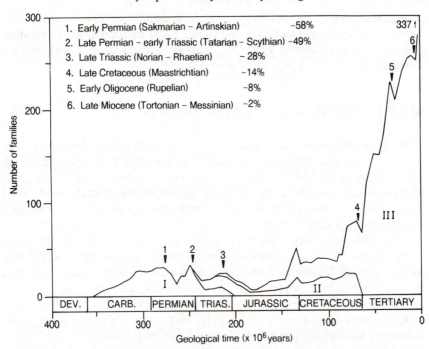

Fig. 4.5 Standing diversity with time for families of terrestrial tetrapods. The upper curve shows total diversity with time. Diversity falls 1–6 represent apparent mass extinctions. I, labyrinthodont amphibians, anapsids, mammal-like reptiles. II, early diapsids, dinosaurs, pterosaurs. III, frogs, salamanders, lizards, snakes, turtles, crocodiles, birds, mammals. After Benton (1987a).

of plant species were gymnosperms, there having been a corresponding decline of pteridophytes, but the total diversity was not much different from the Late Palaeozoic. Following their probable first appearance in the Early Cretaceous, angiosperms increasingly rose to dominance, especially in the Cenozoic, with diversity being much greater in this era than at earlier times, as with marine invertebrates.

The fossil record of non-marine vertebrates is much less good than that of marine organisms, because of problems of preservation and collection bias (Padian and Clemens 1985) but it is clear from Benton's (1987a) analysis that there was a significant increase in family diversity of tetrapods in the Cenozoic following persistently low levels after their Late Palaeozoic appearance (Fig. 4.5). Benton distinguishes three successive family assemblages analogous to Sepkoski's evolutionary faunas, as indicated in the caption to Fig. 4.5. Because of fragmentation of Pangaea the predominant biogeographic theme from Late Mesozoic times onwards was increased provinciality due to isolation, with occasional renewal of land contacts (Padian and Clemens, 1985). The significant Cenozoic diversity increase is likely also to be related to increased latitudinal differentiation due to stronger climatic gradients, similar to the situation with marine invertebrates.

RADIATIONS AND EXTINCTIONS

The most significant radiation was that involving the Metazoa at the beginning of the Phanerozoic, such that by Mid Cambrian times all the existing phyla had emerged (Taylor and Larwood 1990). Subsequent evolution has involved progressive more fine tuning of early established body plans at least as regards invertebrates, bound up with the so-called evolutionary ratchet (Levinton 1988).

The next important radiation involved vascular plants in the Late Silurian and Devonian, following initial colonization of the land (Niklas *et al.* 1985) and associated radiation of terrestrial arthropods, such that a high diversity of insects was established as early as Carboniferous times. Radiation of marine invertebrates following the end-Palaeozoic mass extinctions did not start quickly, and by far the most spectacular Mesozoic marine radiation took place in Mid Cretaceous times, focused on the Albian and Cenomanian stages. The groups that exhibit the radiation most clearly, often even at family level, include planktonic foraminifers, neogastropods, veneroid and hippuritoid bivalves, torticone ammonites, cheilostone bryozoans, decapod and ostracod crustaceans, and teleost fish. On land, the initial radiation of the angiosperms at the expense of the gymnosperms took place at about the same time.

The final major radiation that need be noted here is that of the mammals in the Early Cenozoic, with virtually all the modern orders being established by Eocene times. This involved not only significant diversity increase but size increase within individual lineages and expansion into a wide variety of terrestrial, freshwater and marine aquatic, and aerial niches.

The possible explanations of these radiations is a complex subject beyond the scope of this book and can only be touched upon here. They involve a combination of both biological and physical factors. Once key adaptive innovations had been achieved expansion into new ecological niches was possible. This is perhaps most apparent in the Early Palaeozoic radiation of the Metazoa and Late Palaeozoic radiation of land organisms. For marine benthos the expansion of habitat areas in epicontinental seas as a consequence of eustatic rise of sea level was evidently a major influence, most clearly evidenced for the Cambro–Ordovician and Mid-Cretaceous radiations (Hallam 1992). It had been thought that the relatively sudden appearance of Metazoa in the Early Cambrian was the consequence of skeletonization, but it is clear from trace-fossil evidence that the radiation

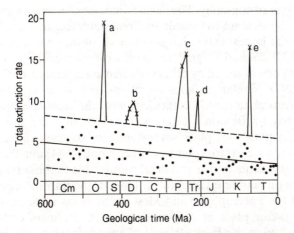

Fig. 4.6 Geological time (Ma)

Fig. 4.6 Extinction rate (families/Ma) of marine animals during the Phanerozoic, showing the 'big five' mass extinctions as peaks with crosses : (a) Late Ordovician; (b) Late Devonian; (c) Late Permian; (d) Late Triassic; (e) Late Cretaceous. Background rates (dots) occur within dashed lines; the solid line is a regression. Cm, Cambrian; O, Ordovician; S, Silurian; D, Devonian; C, Carboniferous; P, Permian; Tr, Triassic; J, Jurassic; K, Cretaceous, and T, Tertiary. Adapted from Raup and Sepkoski (1982).

affected soft-bodied organisms as well. The most plausible explanation for the key environmental trigger, if there was just one, is an increase in atmospheric oxygen concentration beyond a critical threshold.

The subject of mass extinctions has provoked enormous interest in recent years and has been dealt with in a large number of publications, including the following books: Nitecki 1984; Stanley 1986; Donovan 1989; Larwood 1988; Chaloner and Hallam 1989; Eldredge 1991; Raup 1991;. It need only be considered very briefly here.

It is generally accepted that there were five major episodes of mass extinction in the marine realm during the Phanerozoic, which are clearly revealed at family level. These took place at the end of the Ordovician, Permian, Triassic, and Cretaceous and in the Late Devonian. During these events a significant proportion of the marine biota disappeared (Fig. 4.6). In Fig. 4.4 the extinctions are recorded as sharp falls in diversity. By far the biggest is at the end of the Permian. This end-Palaeozoic event does not stand out as a higher peak in Fig. 4.6 because the extinctions are spread out over a longer period of time than the others that show a sharp diversity fall. The five major extinctions are even more clearly recorded at the generic level, together with a greater number of lesser mass extinction events (Sepkoski 1986).

The record of terrestrial life does not reveal such clear-cut results, which is partly due to the poorer preservation potential of the relevant fossils and the less precise stratigraphy. Thus while dinosaurs became extinct at the end of the Cretaceous many mammals and freshwater fish and reptiles survived. As regards the largest extinction event of all, at the end of the Palaeozoic, Bakker (1977) claimed a significant mass extinction event for large reptiles but this has been disputed by others (Padian and Clemens 1985; King 1991; Maxwell 1992). Similarly, claims for an end-Triassic mass extinction event among terrestrial tetrapods coincident with the marine extinctions have been rejected by Benton (1991), who argues that the most important Triassic extinctions took place several millions of years earlier.

With regard to terrestrial plants, neither Knoll (1984) nor Traverse (1988) recognize mass extinction events at all comparable to those that can be recognized among marine invertebrates. Indeed, the transition between what are termed the Palaeophytic and Mesophytic floras is diachronous across the world, taking place at varying times in the Permian of the northern hemisphere and in the Early Triassic of the southern hemisphere.

One interesting question that has emerged is whether or not a clear distinction can be drawn between normal or 'background' extinctions and mass extinctions. Many would argue that mass extinctions are merely the end member of a whole spectrum of events of increasing intensity, but Jablonski (1986, 1989) has made a case, based on a study of Cretaceous molluscan survivorship patterns, suggesting that some traits that tend to confer extinction resistance during times of normal levels of extinction are ineffectual during mass extinction. His hypothesis has biogeographic implications, because it is maintained that, for genera, high species richness and the possession of widespread individual species have imparted extinction resistance during background times but not during the end-Cretaceous mass extinction, when overall distribution of the genus was the important factor.

The causes of mass extinctions are the subject of much controversy. Two schools of thought have emerged, that they are terrestrially or extra-terrestrially induced. The terrestrial phenomena must be global in scope in order to be serious contenders, and the only plausible possibilities involve changes of climate and eustatic sea level. These may of course be interrelated. Thus the growth of polar icecaps as a result of climatic cooling will cause sea level to fall, while a tectonically-induced fall of sea level will lead to greater seasonal extremes of temperature on the more emergent continents, together with greater aridity in the continental interiors. Extinctions due to climatic cooling can reasonably be invoked for Cenozoic and probably a few Palaeozoic extinction events, but less reasonably for the many extinctions that took place when there were no polar icecaps. For these, a better case can be made for marine extinctions bound up with reduction in habitat area of epicontinental seas either as a result of regression due to eustatic fall or the spread of anoxic waters during the early phases of eustatic rise (Hallam 1992).

Extra-terrestrial theories were greatly boosted by the discovery of significant iridium anomalies and shocked minerals at Cretaceous–Tertiary boundary sections across the world, and have led to the extreme view that not only mass extinctions but all extinctions were probably caused by bolide impact (Raup, 1991). However, apart from the Cretaceous–Tertiary boundary the evidence supporting bolide impact is either equivocal, negligible, or non-existent, and even for the celebrated K–T event it is far from clear to what extent the effect of impact, even if accepted, can be separated from that of climatic and eustatic events.

The reason why vascular plants have been comparatively extinction-resistant is probably bound up with the fact that they have evolved numerous mechanisms for coping with increased environmental stress, including long dormancy periods that allow them to survive short-term perturbations (Knoll 1984).

TRENDS WITH GEOGRAPHIC COMPONENTS

One of the most striking geographic trends that has emerged from recent research involves benthic communities in epicontinental seas, with new faunas containing morphological innovations originating in onshore regions and spreading progressively offshore (Jablonski *et al.* 1983; Jablonski and Bottjer 1990*a*, *b*; Sepkoski 1991).

The fullest data come from the Palaeozoic. Sepkoski (1991) analysed over 500 fossil assemblages of known age and environment gleaned from North American palaeontological literature. Figure 4.7 is a time–environment diagram showing the distribution of his so-called evolutionary faunas I, II, and III (Sepkoski 1981 and Fig. 4.4), here termed respectively the Cambrian, Palaeozoic, and Modern faunas. Six environments are distinguished which form an onshore–offshore gradient, from peritidal to deep-water basinal. There is a strong onshore–offshore component to changes between evolutionary faunas, with rapid offshore expansion of the Palaeozoic fauna (and retreat offshore of the Cambrian fauna) during the Ordovician, and slower offshore expansion of the Modern fauna during the later Palaeozoic. The Modern fauna skips to the most offshore zone during the Devonian.

With regard to post-Palaeozoic faunas, it is well known that a number of marine invertebrate groups now largely confined to relatively deep-water habitats, such as stalked crinoids, hexactinellid sponges, and many brachiopods, flourished in shallow-water habitats in the Early Mesozoic. The so-called Mesozoic marine revolution of Vermeij (1977), involving the radiation of durophagous predators, is likely to have made sessile epifauna more vulnerable than hitherto, and migration to a deeper-water refuge would have provided one means of escape. It is thought by Briggs (1974) that most colonizations of abyssal and trench habitats in the deep ocean have probably took place no earlier than the Late Cenozoic.

An intriguing parallel to the onshore–offshore pattern among marine benthic organisms has been proposed for Late Palaeozoic terrestrial vegetation by Dimichele and Aronson (1992). These workers undertook a statistical analysis of 68 floras from the Upper Carboniferous and Lower Permian of Europe and North America. Wetland assemblages are the most commonly encountered floras from the Carboniferous, whereas floras from drier habitats characterize the Permian. Drier habitats appear to be the sites of the first appearance of orders that became prominent in the Late Permian and Mesozoic. These new taxa are thought to have originated in physically heterogeneous, drier habitats, which were geographically marginal in most of the Carboniferous. They subsequently moved into lowlands during periods of climatic drying in the Permian, replacing the older wetland vegetation. Mesozoic wetland clades are derived from Permian dry lowland vegetation.

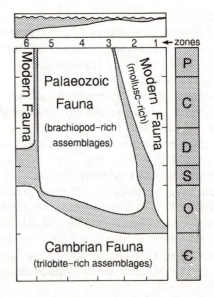

Fig. 4.7 A time-environment diagram illustrating the distribution of evolutionary faunas with respect to marine environments through the Palaeozoic. A generalized environmental gradient, from onshore on the right to offshore on the left, is depicted at the top. This gradient is divided into six environmental zones; 1, nearshore, peritidal habitats; 2, nearshore, protected subtidal environments; 3, offshore, wave-agitated zones; 4, open, mid-shelf environments; 5, deeper, outer-shelf environments; 6, deep-water, off-shelf, and basinal environments. Є, Cambrian; O, Ordovician; S, Silurian; D, Devonian; C, Carboniferous, P, Permian. After Sepkoski (1991).

Vermeij (1987) makes an important distinction between two types of biological refuge. The first type is that of a 'safe house' for adaptively anachronistic forms, where selection pressures are less intense. This is clearly pertinent to the displacement into deeper marine waters of evolutionary older taxa. The second type involves range retraction, due to changing climate and/or geography. This is more relevant to latitudinal distributions that change with time, and to the relative age of tropical and polar taxa, a subject reviewed by Crame (1992a; see also Crame 1993).

Crame distinguishes three hypotheses:

1. Taxa arise in the tropics and disperse to the poles. This is a widely held view dating back to Darwin, Wallace, and Matthew. It seems logical to infer that high-diversity foci were the sites of evolutionary innovation, and studies on marine invertebrates have shown that tropical faunas are apparently characterized by geologically younger taxa.

2. Taxa arise at the poles and migrate to lower latitudes. Just because taxa may have proliferated in a given adaptive zone does not necessarily mean that they originated there. Polar regions experienced less harsh climates in the past and there are indications from the fossil record that a number of plants and invertebrates originated in Antarctica. Some Cretaceous molluscs including the bivalve *Inoceramus* and predatory neogastropods may also have evolved initially in higher latitudes.

3. Polar taxa are remnants of former cosmopolitan distributions. This is a vicariance explanation that eliminates the need for dispersal, but models 1 and 2, though they may be dismissed as mere narrative explanations, are in fact amenable to testing by the fossil record.

THE ROLE OF COMPETITION

What Darwin (1859) had in mind when he discussed the 'struggle for existence' was the eventual success of adaptively superior organisms, a concept that has been uncritically accepted by generations of evolutionary biologists and has received a modern formulation by Van Valen (1973) with his well known Red Queen hypothesis. This type of competition implies that later arrivals on the evolutionary scene can displace the inhabitants of given ecological niches and is hence appropriately termed *displacive competition* (Hallam 1990). On the other hand success may favour the incumbents. The earlier species occupant of a niche would stay there until some physical disturbance caused its elimination. Only after the niche had been so vacated could another species, which could be a direct evolutionary descendant, come to reoccupy it. This alternative phenomenon can be termed *pre-emptive competition* (Hallam 1990) and implies that the prime motor of evolutionary change is, contrary to Darwin's belief, the physical

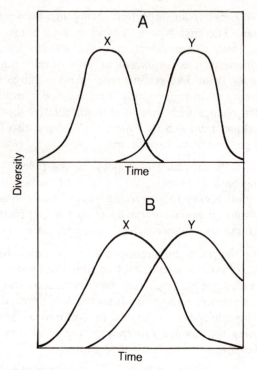

Fig. 4.8 Alternative competition models for patterns of diversity change through time of groups X and Y. A, pre-emptive competition; B, displacive competition. After Hallam (1990).

not the biotic environment. Indeed, evolution could conceivably grind to a halt in the absence of abiotic change, a thought that has led Stenseth and Maynard Smith (1984) to formulate the so-called stationary model of evolution, as opposed to the Red Queen model characterized by continuous biotic interactions.

The rival Red Queen and stationary hypotheses carry different implications regarding changes of faunal diversity with time and can be formulated as testable models (Fig. 4.8). In the pre-emptive model (Fig. 4.8*a*) two groups successively rise to peaks of diversity and then decline, with most of the decline of the older group (X) taking place before the diversity of the younger group (Y) had begun to rise significantly. The implication is that a niche had to be at least substantially vacated before the younger group became dominant. For the displacive model (Fig. 4.8*b*) the stratigraphic ranges of the two diversity curves overlap considerably. This pattern, unlike that of Fig. 4.8*a*, is consistent with a prediction of the displacive competition hypothesis that the rise to dominance of the younger group is the cause of the decline of the older group.

It has become increasingly clear that in general the fossil record of both vertebrates and invertebrates supports the pre-emptive model (Benton 1987*b*; Hallam, 1990). This is most apparent from the record of mass extinctions and subsequent radiations, for example the end-Mesozoic extinction of the dinosaurs and early Cenozoic radiation of the mammals, but is clear also from a host of lesser extinction and radiation events. Rosenzweig and McCord (1991) believe that the most significant phenomenon for long-term evolutionary progress may be the process of what they call *incumbent replacement*. New-clade species acquire a key adaptation that gives them a higher competitive speciation rate than old-clade sources of replacement of extinct species. Incumbent replacement proceeds at a rate limited by the extinction rate and often seems to be linked to mass extinction events.

An interesting parallel to the pattern of purely temporal change discussed above can be made with the geographic changes with time involved in the onshore–offshore patterns of change of benthic invertebrates dealt with earlier. Jablonski and Bottjer (1990*b*) argued that such patterns must have been driven by intrinsic evolutionary processes rather than extrinsic mechanisms, but this has been challenged by Sepkoski (1991).

The process of offshore expansion could involve the competitive superiority of onshore clades resulting from adaptation to stressful or rigorous conditions, or competitive exclusion involving the active displacement by nearshore predators or 'bulldozing' bioturbators. The alternative, which Sepkoski prefers, involves pre-emptive competition and incumbent replacement in which extinction-resistant taxa, as might evolve onshore, hold their ground longer and therefore expand in the face of background extinction among less resistant offshore species. If extinction intensity is highest in nearshore habitats, extinction-resistant clades will expand preferentially in an onshore direction, build up diversity there and then diversify outwards towards the offshore. Thus onshore–offshore patterns of diversification might be the expectation for faunal change quite independently of whether or not clades originate onshore.

Knoll (1984) and Traverse (1988) have maintained that the turnover of plant taxa with time exhibits a fundamentally different pattern from that of animals, and that displacive competition plays a major role. However, Dimichele and Aronson (1992) dispute this for the Carboniferous–Permian vegetational transition, which they consider to be replacive rather than displacive. Taxa that originated in peripheral, drier habitats in the Late Palaeozoic tend to be subgroups of seed plants. The life histories of seed plants suggest *a priori* a greater resistance to extinction than most groups of 'lower' vascular plants, such as ferns, lycopsids, and sphenopsids. This prediction is confirmed by the nearly continuous expansion of seed-plant diversity during the post-Palaeozoic, at the expense of the lower vascular plants (Niklas *et al.* 1985).

None of what has been written in this section should be held to deny a role for biotic competition in evolution. A good case for co-evolution has been made, for instance, by Vermeij (1977, 1987), with the development in the Late Mesozoic of potential prey bivalve and gastropod shells with protective armour *pari passu* with the radiation of durophagous predators. Nor can it be doubted that there was significant co-evolution between insects and angiosperms, but this cannot be established from the fossil record for obvious reasons.

5
Early Palaeozoic

Following the general acceptance of plate tectonics, studies have shown that the disposition of continental areas in the Early Palaeozoic was totally different from that seen today. Based on a combination of biogeographic, palaeoclimatic, and palaeomagnetic data, it is widely believed there was a series of relatively small continents occupying mainly low palaeolatitudes and one large continent, Gondwana. In the east, components of Gondwana including the present Australia and adjacent parts of eastern Asia occupied a low latitude zone, while further west, the present Africa and South America occupied high southern latitudes (Figs 5.1–5.3).

Working with a large number of colleagues Scotese has been a leading figure in producing global reconstructions of continental positions; these have changed over the years, but not in any drastic way. Despite the inevitable limitations, the series of reconstructions presented in McKerrow and Scotese (1990) provides a valuable framework for studying both Early and Late Palaeozoic biogeography.

In the latest set of reconstructions Scotese and McKerrow (1990) distinguish a number of continents that bear little relation to the present array. *Laurentia*, corresponds earlier in the subera substantially to North America. *Baltica* embraces northern Europe bounded on the west by the Iapetus suture, on the east by the Urals suture, on the south by the Variscan suture and on the south-west by the Tornquist Sea. After the Silurian Scandian Orogeny it sutured with Laurentia.

Avalonia comprised England and Wales, south-eastern Ireland and northern France, together with the Avalon Peninsula of eastern Newfoundland and much of Nova Scotia and central New England. It rifted from Gondwana in the Early Ordovician and probably collided with Baltica in the latest Ordovician. The Yucatan Peninsula of Mexico, Florida, the Avalon Peninsula of eastern Newfoundland, central and southern Europe, and a number of Cimmerian terranes were all adjacent to Gondwana at some time in the Palaeozoic. By the end of the Early Palaeozoic Laurentia, Baltica, and Avalonia were sutured together.

With regard to the present components of Asia, *Siberia* and *Kazakhstania* are bounded on the east today by the Verkhayansk fold belt, on the west by the Urals suture and on the south by the South Mongolian Arc. Previous treatments have separated the two areas. Through most of the Palaeozoic

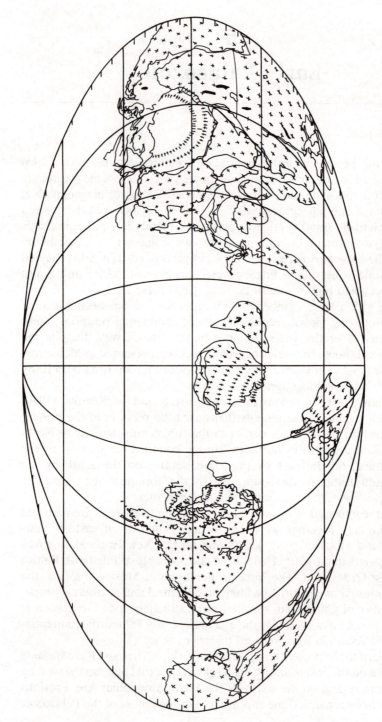

Fig. 5.1 Distribution of continents in the Late Cambrian. After Scotese and McKerrow (1990).

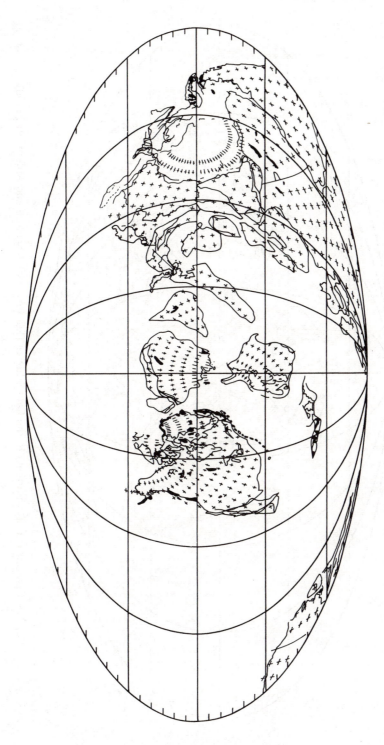

Fig. 5.2 Distribution of continents in the Late Ordovician (Llandeilo–Caradoc). After Scotese and McKerrow (1990).

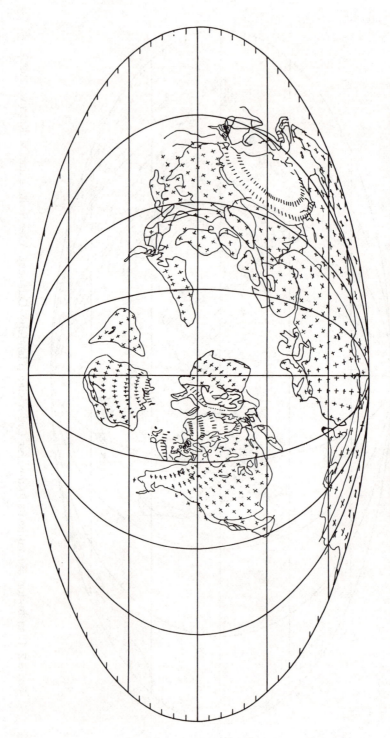

Fig. 5.3 Distribution of continents in the Late Silurian (Ludlow). After Scotese and McKerrow (1990).

they moved northwards, rotating clockwise in the Mid Carboniferous and collided with Baltica along the Urals suture in Late Carboniferous–Early Permian times.

Further east, there are believed to have been three Palaeozoic continents, *North China, South China*, and *Tarim*. In the Scotese and McKerrow (1990) maps Tarim and North China are shown together. Biogeographic affinities suggest that South China was close to the Australian part of Gondwana during most of the Palaeozoic. North China was probably also near eastern Gondwana in the Early Palaeozoic, but the location of North China–Tarim in the Early and Mid-Palaeozoic is less certain than for other continents.

In an exercise in which there are many degrees of freedom it is difficult to have a high measure of confidence in the accuracy of the reconstructions, but they appear to be the best we have. Ideally, one would like to test biogeographic models against reconstructions solidly established by geological and geophysical data, but sometimes the biogeographic evidence is superior to that from other sources, or at least as reliable. There seems to be no escape from persistently attempting to improve upon the most plausible reconstructions using the widest possible array of evidence, and making judgements about the relative merits of different types of evidence in particular cases.

In evaluating plausibility there are two contrasting approaches that should be avoided. One is an unduly cavalier treatment of continent positions with respect to each other solely on the basis of biogeographic data. Thus in the Ordovician reconstruction of Burrett *et al.* (1990) North China is placed near the Shan-Thai region of South-east Asia and Australia, while South China is placed next to the Himalayan–Iranian sector, separated from North China by Tarim and Indo-China. Even more extravagant, Tuckey (1990) proposes, on the basis of a study of Silurian bryozoans, that South China's faunal similarity with Mongolia in Llandovery and Wenlock times indicates that it may have drifted northwards in the Ordovician and Silurian before returning to equatorial latitudes in the Permian.

On the other hand, Boucot has been prepared to accept plate tectonics to the extent of adopting the reconstruction of Wegener's supercontinent Pangaea, but has steadfastly refused to go beyond this, preferring a coherent Pangaea persistent through the Palaeozoic, whereas most geologists accept that it did not come into existence until the Permian (Boucot and Gray 1983; Boucot 1990). This extreme view can only be maintained by a stubborn disregard of a wide array of geological, geophysical, and biological data suggesting a series of oceans narrowing and continents colliding through the course of the era.

Early Palaeozoic biogeography, unlike that for later times, is almost exclusively concerned with marine invertebrates. The global biogeography of the Cambrian, Ordovician, and Silurian periods will be dealt with successively before more detailed attention is paid to changes through the subera in two important and well-studied regions. These concern firstly the countries presently situated adjacent to the North Atlantic and secondly the relationship of East Asian continental areas, mainly China, to other continents.

CAMBRIAN

Good biogeographic information is much more limited for the Cambrian than the rest of the Early Palaeozoic, being confined essentially to trilobites.

Palmer (1973) was the first to establish clearly that the trilobite faunas of continental cratonic interiors were dominated by endemic taxa while those in the deeper water facies of the craton peripheries contained a significant proportion of cosmopolitan taxa. These cosmopolitan trilobites include eodiscids, oryctocephalids, and pagetids in the Lower and Middle Cambrian and agnostids in the Middle and Upper Cambrian. The open ocean was perceived as a genetic reservoir. A series of extinctions, defining so-called *biomes*, took place among the cratonic faunas, and new faunas were derived from oceanic sources.

It is evident from this that the cratonic trilobites are the only ones that are biogeographically useful. For the Early Cambrian Palmer distinguished two faunas, the olenellid and redlichiid. The olenellids were confined to North and South America and the north-western extremities of Europe, while the redlichiids are characteristic of China, South-east Asia, Australia, and Antarctica, but also occur in the Mediterranean region. For the Mid and Late Cambrian a western European 'province' characterized by olenids, conocoryphids, and paradoxids is distinguishable from a North American 'province' with oryctocephalids and several other families. A third 'province' in South-east Asia and Australia is characterized especially by damesellids. There are striking similarities between faunas of all ages in Antarctica with those of Siberia and other areas bordering the West Pacific. The faunas of the Andean pre- cordilleran region in Argentina are very similar to those of western North America.

A more detailed study of Middle Cambrian trilobites, using Simpson similarity coefficients with cluster analysis, was undertaken by Jell (1974). There was no analysis for biofacies, but it is considered that the biogeographic influence over similarity was greater than the biofacies influence. Thus eastern Newfoundland and the neighbouring eastern North American seaboard areas are characterized by siliciclastic sediments as in the northern Appalachians but the faunas of the two areas are dissimilar, with the former having European affinities. Pelagic miomerid genera (13 per cent of the total) are cosmopolitan and were in consequence excluded from the analysis.

Jell distinguishes three provinces, between which only a relatively small number of genera and even families migrated:

1. *Columban*; embracing North America (excluding the easternmost part) and south America. This was the most distinct and evidently most isolated.

2. *Viking*; including Europe, Turkey, Israel, the Atlas Mountains of Morocco, and eastern maritime North America. At the generic level it was entirely distinct from the Columban Province, but there was some interchange with Siberia.

3. *Tollchuticook*; including Siberia, China, Australia, and Antarctica. This corresponds to the Redlichiid Realm of other authors. *Redlichia* also occurs in Spain, but the provincial affinities of the total Spanish fauna are totally Viking. Therefore Jell considers that *Redlichia* should not be used to define a province.

The agnostids transgress the provincial boundaries.

The main change through the course of the Mid Cambrian was an increase in interchange between Siberia and Baltica and decrease between Siberia and southern Europe. By using palaeomagnetic data Jell attempted a reconstruction of continents which is not strikingly different in most respects from those of Scotese (Jell 1974, Fig. 10). One obvious feature of the reconstruction is that most strata with Middle Cambrian trilobite faunas lie within 30° of the palaeoequator, suggesting that latitudinal influences on the faunas were minimal. It is thought that barriers to migration must have been either land or wide stretches of deep ocean.

A feature of particular interest is that the Atlas Mountains of Morocco are characterized by thick carbonates and, in the Lower Cambrian, prolific archaeocyathid faunas, strongly suggesting to Jell a low-latitude site adjacent to Europe rather than Africa, placed at a high latitude in his reconstruction.

However, in the alternative Middle Cambrian reconstruction of Scotese and McKerrow (1990, Fig. 4) North Africa is placed at a low latitude adjacent to South America. Southern Europe is at a higher latitude and Baltica far distant at an even higher southerly latitude. The gross disparity in this respect with Jell's reconstruction is clearly a reflection of our ignorance.

Jell's faunal provinces have been utilized by Dalziel (1992) in an unusual reconstruction of the Gondwana supercontinent at its final amalgamation in the Cambrian (Fig. 5.4). It is unusual in that North America has been placed opposite South America across the newly opened Iapetus Ocean. Even allowing for the different projection, this juxtaposition is

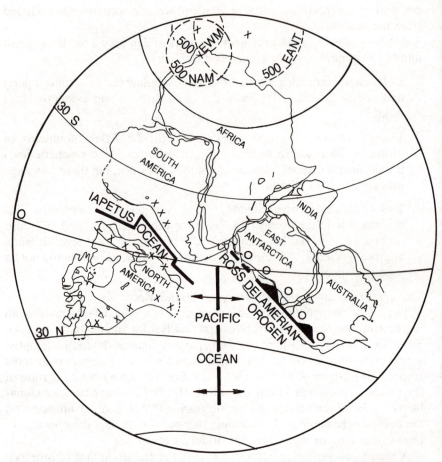

Fig. 5.4 Palaeomagnetically controlled reconstruction of the Pacific and southern Iapetus oceans on final amalgamation of the Gondwana Supercontinent at the end of the Cambrian. Crosses and circles respectively denote the North American and East Antarctic/Australian trilobite faunal realms of Jell (1974). After Dalziel (1992).

clearly different from the relative positions of these continents in the Scotese reconstructions, though both satisfy the latitudinal constraints provided by palaeomagnetic data, and perhaps should serve as a further reminder of the difficulty in fixing accurately continental positions so far back in time.

The only other fossil group in the Cambrian to have received biogeographic attention is the archaeocyaths or archaeocyathids, which flourished in the Early Cambrian and disappeared soon afterwards. They were the first reef invertebrates, and thrived in the shallow waters on the flanks of carbonate belts. Their lack of spicules contrasts with non-calcareous sponges but their sponge-like organization suggests that they are best regarded as an extinct class of the Porifera. For the later part of the Early Cambrian Zhuravlev (1986) recognises two provinces which must have occupied low-latitude zones. The American–Koryakiyan Province formed in western North America (from Alaska to Sonora) and the Afro–Siberian–Antarctic Province probably extended from North Africa and western Europe to Siberia, Australia, and Antarctica.

ORDOVICIAN

The groups that have been most thoroughly studied from a biogeographic point of view are the trilobites and graptolites; accordingly these warrant special treatment.

Trilobites.
The first attempt to analyse trilobite provinciality after the plate tectonics revolution was by Whittington and Hughes (1972). For the Early Ordovician they distinguished four provinces, the Asaphid, Bathyurid, Asaphopsis, and Selenopeltis, named after characteristic taxa. They were believed to have lived in shallow waters around individual continental blocks. The oceans that separated these blocks were barriers to migration so that evolution took place independently in each province.

However, on the basis of a detailed study of Early Ordovician trilobites and associated facies in Spitsbergen, Fortey (1975) clearly demonstrated the importance of taking communities into account before generalizing about faunal provinces. Fortey's studies led him to infer a shallow to deep water environmental gradient, characterized respectively by illaenids and cheirurids, nileids, and olenids (Fig. 5.5). The illaenid and cheirurid community corresponds to typical Bathyurid Province, but the nileid community contains a fauna in which almost every genus can be matched with the Asaphid Province. The olenid community has counterparts in Bolivia and Peru belonging to the Asaphopsis Province. The Spitsbergen strata thus contain faunas suggestive of connections with three of the four provinces of Whittington and Hughes, so the oceans could not have been such effective barriers to communication as claimed.

The olenid community, evidently a deep-water one often associated with graptolitic shales (Fig. 5.5), is the most cosmopolitan, and evidently compares with the Cambrian agnostids.

Graptolites.
Because of its planktonic habitat and pelagic distribution, this group is more cosmopolitan than benthic groups such as trilobites and hence

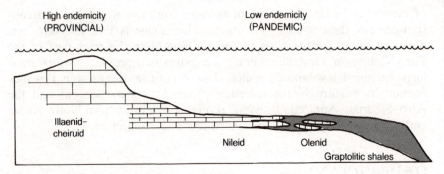

Fig. 5.5 Palaeogeographic model showing disposition of community types across an Early Ordovician epicontinental sea. Adapted from Fortey (1975).

less valuable for locating continental positions. Traditionally two rather unfortunately named realms or provinces have been distinguished, the *Atlantic*, characteristic of Europe, and the *Pacific*, characteristic of North America and Australia, with the degree of provincialism diminishing in the later Ordovician (Skevington 1974; Berry and Wilde 1990; Finney and Chen 1990). The Pacific faunas are characterized by many species of *Isograptus*, and *Oncograptus* and *Cardiograptus* are endemic. The Atlantic faunas in the Lower Ordovician include abundant didymograptids with pendent stipes.

Several different environmental interpretations have been proposed to account for the provinciality.

1. Latitudinal surface-water temperature gradients. Skevington (1974) indicated that in palaeomagnetically-controlled reconstructions, Pacific faunas occupied low and Atlantic faunas higher latitudes. Corresponding to this, diversity of the Pacific faunas is higher. Reduction of provinciality in the Mid and Late Ordovician was attributed by Skevington to climatic cooling, because the graptolites could not survive at high latitudes and became restricted to the tropics, such that a globally rather homogeneous fauna became established towards the end of the period. However, this purported extinction of graptolite species in the Late Ordovician is contradicted by the existence of widespread assemblages in North Africa (Cooper *et al.* 1991). Another difficulty of Skevington's model is that Atlantic as well as Pacific faunas occur in China, which according to the Scotese reconstructions lay in low latitudes (Finney and Chen 1990).

2. The temperature-related partitioning of epicontinental and oceanic water masses (Finney 1984, 1986; Finney and Chen 1990). This interpretation leads to some curious and implausible conclusions. It is argued that the occurrence of Pacific faunas in Argentina can be explained by the existence of a large Pacific gyre, but this could not explain the existence of Atlantic faunas in neighbouring Bolivia and Peru. An *ad hoc* clockwise rotation of South America is therefore invoked to bring Argentina more into the tropical zone and Bolivia and Peru to higher latitudes. The inboard position of the South China region within the tropical zone allowed for the establishment of a warm environment adjacent to but free of a cold, north-east flowing current that bathed the Central China region (Finney and Chen 1990).

3.*Varied depth distributions.* Cooper *et al.* (1991) distinguish three depth-related environments recognizable by characteristic sedimentary associations:

(a) shelf and platform (0–200 m); shallow-water carbonates and silici-clastics;
(b) outer shelf-upper slope (~200–2000 m); mudstones and proximal turbidites;
(c) lower slope–ocean floor (>2000 m); pelagic, organic-rich and siliceous mudstone, chert, turbidites, debris flows. Of particular importance are the regions where the depth profile ranges from shallow shelf to offshore basin, because in this case any faunal differences cannot be attributed to latitudinal changes.

Three biotopes are distinguished (Fig. 5.6). Graptolites not restricted to particular depth facies are inferred to belong to the *epipelagic biotope*, which probably embraces most Ordovician species, and is characterized especially by the phyllograptids and diplograptids.

Within this biotope pandemic, Atlantic and Pacific groups can be distinguished. The largest, pandemic, group must have been eurythermic. The *deep-water biotope* is characterized especially by isograptids and sinograptids, with both Atlantic and Pacific 'provinces' having representatives. Finney (1984, 1986) has argued that this biotope signified an open ocean water mass. Cooper *et al.* (1991) put forward four arguments against this idea. There is no 'leakage' of oceanic species into shelf waters, as occurs with Recent plankton. During marine transgressions the deep-water biotope fauna encroached on to the outer shelf, showing that the shelf edge itself did not mark the position of the biotope boundary. A large segment of the plankton—the epipelagic biotope fauna—was clearly not confined by the shelf–ocean boundary, suggesting that this boundary had little effect on graptolite distribution. Fragile, multistipe forms, poorly designed for survival in shallow environments, are confined to the deep-water biotope.

The *inshore biotope*, characterized by pendent didymograptids, may be an exception to the above rule. It was apparently confined to epicontinental

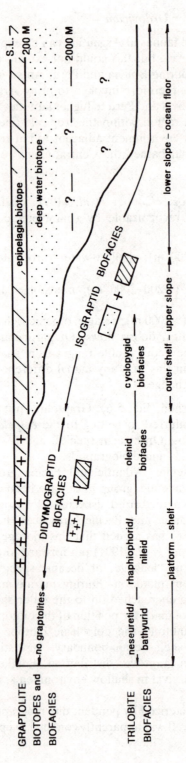

Fig. 5.6 Inferred distribution of Early Ordovician graptolite biofacies, from the inner shelf to the ocean floor, in relation to contemporary trilobite biofacies. Adapted from Cooper et al. (1991).

waters because of the physical properties of the water mass rather than depth.

The model of Cooper *et al.* (1991) seems to account for the biogeographic distribution of graptolites better than any other.

While there is a well-established relationship between marine transgressions and radiations of benthic invertebrates, and between regressions and extinctions (Hallam 1992) it is surprising to find a similar relationship claimed for planktonic graptolites (Berry and Wilde 1990). According to these authors, graptolites are likely to have lived in high productivity, oxygen-poor, nitrogen compound-rich waters, as characterized today by both low latitudes and upwelling zones in higher latitudes. If organic productivity were great enough to create a condition whereby such waters could rise to within a few metres of the ocean surface, even epicontinental seas could have been sites for the deposition of graptolitic shales. Sea- level rise could have led to transgression of oxygen-poor waters while regressions could have led to the draining of such waters from the shelves.

Other groups.

Brachiopods compare with the great majority of trilobites (excluding olenids) in being essentially neritic organisms exhibiting relatively high levels of endemism but they do not appear to exhibit any provincial patterns that cannot better be inferred from trilobites (Cocks and Fortey 1990). An especially interesting story has been worked out for the latest Ordovician by Sheehan and Coorough (1990). A distinctive *Hirnantia* fauna became widely established coincident with a major glacioeustatic regression induced by the growth of a Gondwana icecap. Elements of this fauna first appeared in high southern palaeolatitudes and expanded rapidly northwards, as far as Sweden, as the temperature fell. Along with many other invertebrates it then became extinct at the end of the period.

An explosive radiation of conodonts in the Early Ordovician gave rise to the group's greatest Phanerozoic diversity (Bergström 1990). In the Late Cambrian, two faunal regions can be distinguished, the Midcontinent, embracing the North American interior, Siberia, Australia, North China, and the Atlantic, embracing European and some Asian countries. Faunas of Baltica and South Central China are closely similar to each other but strikingly different from those of the Midcontinent faunal region. Baltic and British faunas are similar in several respects but a number of important differences justify separation of British and Baltic provinces within the Atlantic region.

Climate is considered to be the most important controlling factor, with the Atlantic faunas signifying high latitude and cold water. However, Baltic conodonts are known from marginal areas of the North American plate but absent from the North American and Siberian cratonic interiors, and Atlantic conodonts are typically found in deeper water deposits. It would appear more reasonable, taking into account Fortey's work on trilobites, to suggest that the Atlantic faunas belong to a more pelagic, cosmopolitan biotope. This would help to explain the striking similarities between Europe and China. Further analysis of facies associations and degrees of endemicity seems to be desirable for this group before definitive statements are made about provinciality.

The biogeography of Ordovician nautiloids has been analysed by Crick (1990) using the probabilistic index of similarity of Raup and Crick (1979). Palaeozoic nautiloid distributions indicate that they were not part of the nekton capable of oceanic dispersal, but shelf benthos subject to little or no post-mortal transport.

As with the conodonts, there was a large increase in diversity at the start of the period, in the Tremadoc. The more ancestral ellesmeroceratids dominated the faunas of North China and Laurentia, while the newly radiating endoceratids occurred on all landmasses, showing faunal connections between Kazakhstania, Siberia, Baltica, and northern Gondwana. Most of the faunas of Baltica, Kazakhstania, and the South American fauna of Gondwana are significantly dissimilar to those of Laurentia, indicating faunal isolation. The significantly similar faunas of Siberia and Laurentia are less difficult to explain than the significant similarity between the eastern Australia fauna of Gondwana and Alaska. Significant dissimilarity between Kazakhstania and North China suggests the existence of a barrier to communication. The faunas with the highest diversity and where a few genera could establish connections over large distances were located in low latitude regions.

Generic diversity peaked in the Arenig, with an endemicity level comparable with the Tremadoc. Endoceratids, actinoceratids, and tarphyceratids were the least endemic groups; ellesmeroceratids and orthoceratids were largely endemic to Laurentia. The Llanvirn marks the peak isolation of Laurentia and increasing similarity between Siberia, North and South China, and the proximal parts of Gondwana. In the later Ordovician there was a trend towards increase in significantly similar faunas among most landmasses and a corresponding decrease in significantly dissimilar faunas.

Cratonic and pelagic biogeography

Cocks and Fortey (1990) distinguish four cratonic faunas in the Early Ordovician (Arenig–Llanvirn; Fig. 5.7).

1. *Gondwana* faunas have highly characteristic trilobites and brachiopods in inshore sites. The trilobites include certain Calymenacea and Dalmanitacea and these names are used to define the province, rather than the Selenopeltis Province of Whittington and Hughes (1972),

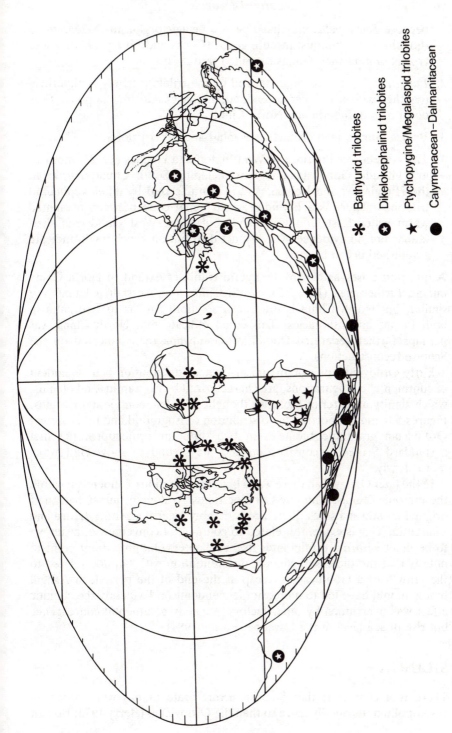

Fig. 5.7 Distribution of Early Ordovician (Arenig–Llanvirn) platform trilobite assemblages. After Cocks and Fortey (1990).

Bathyurid trilobites ✳

Dikelokephalinid trilobites ✪

Ptychopygine/Megalaspid trilobites ★

Calymenacean–Dalmanitacean ●

because *Selenopeltis* may have been a pelagic organism. *Neseuretus* is thought to be the most inshore genus (Cocks and Fortey 1988). These typical Gondwanan faunas extended into Avalonia.

2. *Laurentia* faunas are characterized by the trilobite family Bathyuridae together with a number of endemic molluscs and brachiopods. They occur also in Siberia and North China.

3. *Baltica* faunas have distinctive asaphid and other trilobites.

4. The Asaphopsis Province of Whittington and Hughes (1972) embraces low latitude Gondwanan regions including South China, Australia, the Himalayas, and Bolivia, and is characterized by dikelokephalinid trilobites. There is a latitudinal and faunal cline between presumed warm water East Gondwana and cool water West Gondwana shelf faunas but no evidence of a disjunct distribution which might indicate a vanished ocean barrier.

A question arises as to why bathyurid trilobites extend to North China but no further east (Fig. 5.7). South China and Australian faunas are similar, but North China is different and more similar to Laurentia in both faunas and lithofacies. This could indicate that North China was perhaps further separated from East Gondwana than indicated on the Scotese reconstructions.

Early Ordovician pelagic organisms, whose distribution is independent of continental configurations, are characterized by isograptids and olenids, which signify an exterior biofacies thought to define continental margins. Figures 5.8 and 5.9 portray the distribution of isograptid and other Lower Ordovician graptolite biofacies in two different projections, the first a standard Scotese reconstruction and the second to show Gondwana more clearly.

In the later Ordovician there was a gradual reduction of endemism within the cratonic faunas. Thus by Caradoc times Avalonian faunas had their original Gondwana stock diluted considerably by arrivals from Baltica and Laurentia. This is evidently related to continental convergence, a subject to be dealt with more fully later in this chapter. One important point to note is that the increased climatic differentiation with latitude related to the growth of a Gondwanan icecap at the end of the period, in Ashgill times, should have led to more, not less endemism. Evidently the climatic effect was overridden by other factors, not only continental convergence but rise of sea level in the Caradoc (Fortey 1984).

SILURIAN

There is a consensus that Silurian invertebrate faunas were relatively cosmopolitan, decidedly more so than the Ordovician (Berry 1973; Boucot

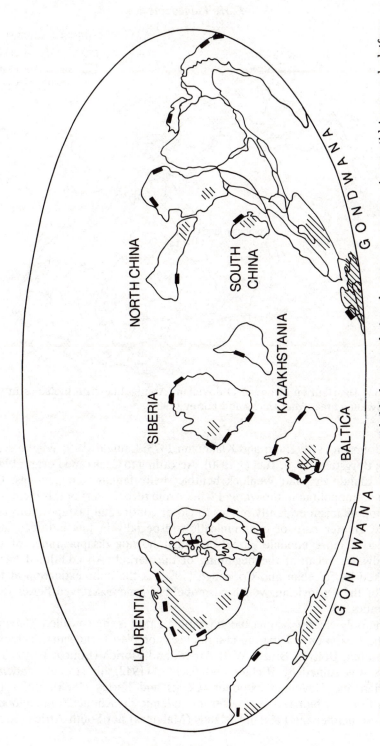

Fig. 5.8 Distribution of Lower Ordovician isograptid (black rectangles) and contemporary non-isograptid immersed platform graptolite biofacies (diagonal shading). After Fortey and Cocks (1986).

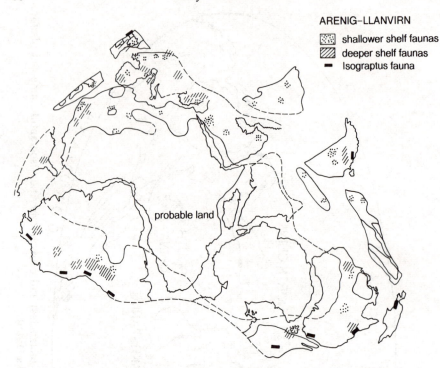

ARENIG–LLANVIRN

░ shallower shelf faunas
▨ deeper shelf faunas
▬ Isograptus fauna

probable land

Fig. 5.9 Distribution of Lower Ordovician (Arenig–Llanvirn) biofacies in the Gondwanan area. After Cocks and Fortey (1988).

and Johnson 1973; Kaljo and Klaamann 1973; Laufeld 1979; Witzke *et al.* 1979; Bergström 1990; Tuckey 1990). According to Cocks and Fortey (1990) Late Llandovery and Wenlock benthic shelly faunas were amongst the most cosmopolitan in the whole Phanerozoic record. Part of this increased cosmopolitanism evidently relates to closer continental juxtapositions and hence greater ease of communication, especially in low latitudes, and part to a more equable global climate following disappearance of the Gondwana icecap at the beginning of the period. An additional factor invoked by Sheehan and Coorough (1990) is the mass extinction at the end of the Ordovician, which cosmopolitan genera survived better than endemics.

The only clearly distinct biogeographic entity is the so-called Malvinokaffric Realm embracing the marginal seas of West Gondwana, including Argentina, Bolivia, Brazil, West Africa, and Florida (Boucot 1990). The name was coined by Richter and Richter (1942) for strongly endemic Silurian and Devonian trilobites. Cocks and Fortey (1988) prefer the term *Clarkeia* fauna, named after an endemic rhynchonellid brachiopod, because neither the Falkland Islands (Malvinas) nor South Africa (Kaffir country) has fossiliferous Silurian strata (Fig. 5.10). The invertebrate fauna is of low diversity and the lack of limestones and reef organisms such as corals and stromatoporoids is indicative of a cold or at least cool climate, which is consistent with the high palaeolatitude position.

With regard to the rest of the world, there is no consensus about biogeographic subdivisions. Boucot (1975, 1990) distinguishes a North Silurian Realm divided into North Atlantic and Uralia–Cordilleran Regions, characterized by high diversity faunas and abundant carbonate facies. The low palaeolatitude position is also consistent with a warm climate. It should be observed here that Boucot's regions do not seem to make much sense on either a Scotese or Pangaea reconstruction.

Differences of opinion exist about changes in the amount of provincialism towards the end of the period. According to Boucot and Johnson (1973) a low degree of provincialism among the brachiopods became established in

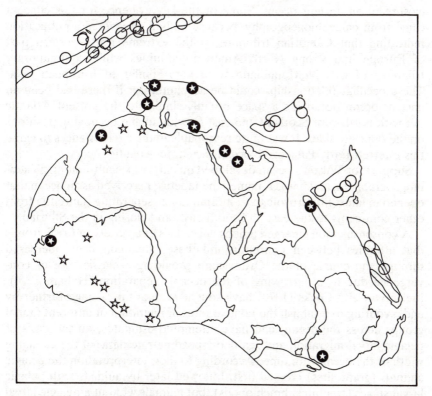

Fig. 5.10 Contrasts in Mid and Late Silurian Gondwanan faunas. Open stars are *Clarkeia* brachiopod faunas of South America and comparable low-diversity faunas in Africa. Circles are diverse low-latitude faunas. Black circles with stars are lower-diversity temperate faunas. After Cocks and Fortey (1988).

Late Wenlock times after a period of almost complete cosmopolitanism, and continued to the end of the Silurian. Analysis of stromatoporoids also showed that the strongest evidence of provincialism appeared in the Ludlow, with clear differences emerging between European and Asiatic faunas (Nestor 1990). On the other hand Kaljo and Klaamann (1973) claimed that by the end of the Silurian provincial differences between coral faunas decreased.

CONVERGENCE OF LAURENTIA, AVALONIA, AND BALTICA

In a seminal paper published shortly before the plate tectonics revolution Wilson (1966) proposed that the Early Palaeozoic Caledonian orogenic belt of north-west Europe and eastern North America marks the line of closure of an ancient ocean. Some of the key evidence for this concept came from palaeobiogeography, because he had discovered a publication indicating that Cambrian trilobites of the extreme north-western part of Europe had strong North American affinities while contemporary trilobites of east Newfoundland were very similar to European taxa. These peculiar relationships could be accounted for if there had been an ancient ocean occupying a space broadly similar to the present Atlantic but with north-west Scotland and east Newfoundland occupying positions on the opposite sides. It was quite reasonable in the circumstances to name this inferred Early Palaeozoic ocean the Proto-Atlantic.

Support for Wilson's ocean developed quickly, especially among palaeo-biogeographers, and it was renamed the Iapetus, perceived as an ocean that opened in latest Proterozoic to Cambrian times, separating Laurentia from other continents to the east and south-east, and closing in the Silurian.

A consensus soon emerged among specialists of a variety of fossil groups that affinities between Laurentia and these other continents increased through the course of the Ordovician, providing a classic case of con-vergence due to a narrowing of the oceanic separation (Hallam 1974). McKerrow and Cocks (1976) have sought to take this a stage further by endeavouring to establish the relative rates of migration of different faunal groups across the ocean, and have attempted estimates, on the basis of pelagic larval migration and sea-floor speading rate data, of the changing width of the ocean with time. According to their interpretation the pelagic animals (graptolites) crossed first, followed later by animals with pelagic larval stages (trilobites, brachiopods), but animals without a pelagic larval stage (benthic ostracods) were not able to cross until the ocean had closed at one point, though not necessarily everywhere along its length. Finally, faunas limited to freshwater or brackish water (like many Siluro–Devonian fish) did not cross until there were non-marine connections between the

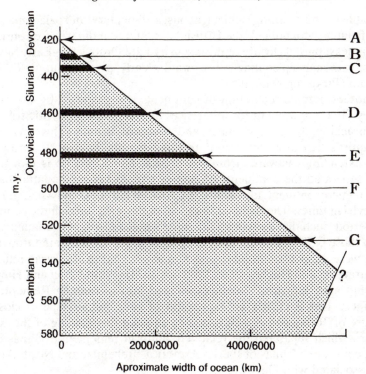

Fig. 5.11 The times at which faunas became common to both sides of the Iapetus Ocean. A, Closure of ocean (Norway); B, freshwater fish; C, benthic ostracods; D, trilobite and brachiopod species; E, trilobite and brachiopod genera; F, *Didymograptus bifidus*; G, *Dictyonema*. After McKerrow and Cocks (1976).

continents on either side of the closing ocean (Fig. 5.11). This took place initially in the north-east, with collision of Scandinavia and Greenland.

As regards thelodont fish, Turner and Tarling (1982) provide support for McKerrow and Cocks' model by pointing out that the strong similarities between taxa of western Russia and the Canadian Arctic are consistent with a Silurian closing in the north. On the other hand Schallreuter and Siveter (1985) observed that ostracods actually matched the story inferred from trilobites and brachiopods in indicating progressively increasing faunal connections between the Laurentia and Baltica/Gondwanan continents through the course of the Ordovician. Because extant ostracods have no pelagic larval stage, Schallreuter and Siveter (1985) thought that the Iapetus Ocean was much narrower than envisaged by McKerrow and Cocks, and probably studded with islands to allow 'island hopping'. Such an idea is supported apparently by a wide range of geological research, because the available tectonic, stratigraphic, igneous, sedimentary, and palaeomagnetic data from Newfoundland, the British Isles, and Scandinavia are consistent with at least partial closure of the ocean by Late Ordovician–Early Silurian times, although epicontinental seaways clearly persisted until the Late Silurian (Pickering *et al.* 1988).

Another oceanic separation of continents was proposed by Cocks and Fortey (1982) to account for differences between shallow neritic trilobites and brachiopods of Scandinavia and southern Britain. This was called Tornquist's Sea after the major tectonic feature known as Tornquist's Line extending eastwards from the present North Sea. It is thought to have separated the continents of Baltica and Avalonia, the latter having been a microcontinent adjacent to the African part of Gondwana, in Early Ordovician times. The faunal differences diminished through the course of the period, such that by Ashgill times the whole of Britain had faunas similar to those of Scandinavia, even at species levels. By the Late Ordovician there were many taxa in the British faunas with North American affinities (Whittington and Hughes 1972; Williams 1973). Whittington and Hughes' Asaphid and Bathyurid Provinces merged into a common Remopleurid Province. This is clearly an expression of concomitant Iapetus closure. Closure of Tornquist's Sea of necessity entailed the opening of the Rheic Ocean, which separated the relatively uniform remopleurid faunas from relict cool water faunas of Iberia–Armorica, Bohemia, and North Africa, still associated with Gondwana.

The ostracod data presented by Vannier *et al.* (1989) provide strong support for Cocks and Fortey's model, not only the convergence with Baltica and Laurentia but divergence between as the Rheic Ocean opened. Thus from Llanvirn–Llandeilo times and through the Late Ordovician the percentage of palaeocope genera common to the British Isles and Iberia–Armorica shows a distinct overall fall.

The existence of Tornquist's Sea was denied by Paris and Robardet (1992) on the basis of their analysis of acritarchs. Fortey and Mellish (1992) responded by pointing out that acritarchs, as a planktonic group, are unsuitable material for such a study. By using both single linkage cluster analysis and parsimony analysis of endemicity, which give similar results, Fortey and Mellish demonstrated a clear three-fold division in the Early Ordovician between Laurentian, Baltic, and 'Gondwanan' faunas among trilobites and ostracods and a slightly less clear division among brachiopods. Conodonts were excluded from the analysis because of their virtual absence from Gondwanan faunas. On the other hand, poor biogeographic discrimination is shown by the acritarchs, as might be expected because of their mode of life.

A review of the latest palaeomagnetic data, especially for the Ordovician, provides support for the concept of Baltica as an individual continental unit in the Early Palaeozoic positioned in high southerly latitudes in an 'inverted' geographic orientation, separated from the northern margins of Gondwana by Tornquist's Sea and from Laurentia by the Iapetus Ocean (Torsvik *et al.* 1992). While undergoing a counter-clockwise rotation Baltica migrated northwards through most of the Palaeozoic, except for a brief period of southerly movement in Late Silurian/Early Devonian times after collision with Laurentia. It is difficult to be precise about when eastern Avalonia collided with the south-western margin of Baltica. Avalonia had clearly rifted off Gondwana by Llanvirn–Llandeilo times and may have collided with Baltica in the Late Ordovician, though presently available Silurian palaeomagnetic data suggest a Late Silurian collision. Baltica and Laurentia collided in Mid Silurian times, with the first collision between Norway and Greenland/Scotland giving rise to the Scandian Orogeny in south west-Norway.

One important aspect of geological history that should not be overlooked in attempting to account for Late Ordovician faunal convergence is the pronounced sea-level rise in the Caradoc (Fortey 1984), with a concomitant increase in diversity and spread of shelf faunas, as acknowledged both by Cocks and Fortey (1982) and Vannier *et al.* (1989).

THE RELATIONSHIP OF CHINA TO OTHER CONTINENTS

Although a great deal of both palaeobiogeographic and palaeomagnetic research has been carried out in recent years the story emerging from China, Australia, and neighbouring countries remains much less clear than for Baltica and adjacent continents.

Ordovician faunal differences between North and South China led Burrett (1973, 1974) to suggest that two continental blocks remained separate until they collided to form the Qinling Suture during the course of the Triassic Indosinian orogeny; supporting evidence was provided from palaeomagnetic results. The case was presented for a much enlarged Early Palaeozoic Gondwana primarily on the basis of the shallow-water faunas of the Shan-Thai Terrane having a remarkably close resemblance to those of the Australian craton.

According to Burrett *et al.* (1990) any reconstruction has to explain why at certain times there are very strong faunal differences between North and South China while at others the faunas are similar. Thus in groups such as nautiloids and trilobites there are close relationships in the Cambrian but almost total dissimilarities in the Ordovician. Taking into account palaeomagnetic data, an anticlockwise movement of Australia, and hence of Gondwana as a whole, is invoked, from the Early Cambrian to the Early Ordovician. A major consequence is that a South China Terrane is brought into progressively higher latitudes. This is held to explain the

Early Cambrian faunal similarities (all terranes subtropical) together with differences between South China (cool waters) and North China (tropical) and faunal similarities between North China, Australia, and Shan-Thai (all tropical) in the Early Ordovician.

There appear to be a number of difficulties with this model, apart from the extravagant geographic changes proposed. According to Crick (1990) there are no shared nautiloid taxa between North and South China in the Late Cambrian but in the Mid Ordovician the two regions form a single biogeographic entity. This is in sharp contradiction to Burrett's claims. With regard to Early Ordovician trilobites, there are indeed marked differences between the North and South China faunas but the latter are strikingly similar to those in Australia, while the former are Laurentian–Siberian in affinity (Cocks and Fortey 1988). The Ordovician corals of North China are likewise most closely related to those of Laurentia and Siberia, while South China was an independent province, though the fauna had close affinities with that in eastern Australia in the Early Silurian (Liao 1990).

Graptolites and conodonts are less reliable for establishing relative continental positions, for reasons mentioned earlier, but Fortey and Cocks (1986) make an interesting point about graptolites in China, where both 'Pacific' and 'Atlantic' faunas occur (Finney and Chen 1990). On modern palaeogeographic maps North and South China are usually shown as split along the Qinling line, a body of faunal discontinuity. If this really represents a closed ocean, it is difficult to account for the absence of the isograptid biofacies, which is thought to be a useful means of distinguishing between deep intracratonic basins and narrow but genuine oceanic basins. Possibly then North and South China were only separated by some deep intracratonic basin. However, in a later paper Cocks and Fortey (1990) raise the possibility that, on grounds of the presence of bathyurid trilobites only in North China, this continental region was more distantly separated from South China than shown on Scotese-type reconstructions.

Whatever the situation was earlier, it is evident that in Early and Mid Silurian times both North and South China share in the general global cosmopolitanism, at least as regards brachiopods, trilobites, and rugose corals. Thus South China has no endemic trilobite genera (Wang *et al.* 1984). More generally, this Silurian cosmopolitanism suggests that no continental regions, containing marine cratonic faunas, were widely separated from each other.

6

Late Palaeozoic

As a result of the so-called 'conquest of the land', the Late Palaeozoic was the time when a substantial biota of vascular plants and vertebrates became established on the continents, thereby expanding the range of organisms whose biogeography can be studied. As regards plate tectonics, the supercontinent Pangaea had been created in the Permian, as a consequence of successive coalescence of continental blocks.

Reconstruction of Pangaea for Permian times has proved problematic, because of an apparent disparity of geological and palaeomagnetic results. When plotted on a conventional Pangaea reconstruction the palaeomagnetic poles for Gondwana are clearly distinct from those of the other major component, Laurussia, consisting of North America, Greenland, and Europe. Attempts have been made to avoid unacceptable overlap of these components by invoking right-lateral shear movement on a huge scale in Late Permian and Triassic times. Morel and Irving (1981) proposed a pre-shear reconstruction in which the northern margins of South America lay adjacent to eastern North America and West Africa against southern Europe. Even more extreme, Smith *et al.* (1981) had South America against Europe and Africa against southern Asia (Fig. 6.1).

Hallam (1983*b*) dismissed these reconstructions on a variety of geological and paleobiogeographic grounds, and his paper provoked a re-evaluation of the palaeomagnetic data (Smith and Livermore 1991). It appears that the best match still implies a large displacement between Laurussia and Gondwana but there are other statistically acceptable matches that require much smaller displacements. Palaeomagnetism alone cannot discriminate between the various possibilities and the geological evidence favours Pangaeas with small displacements. It is best therefore to adhere to conventional Pangaea reconstructions, as is done by Scotese and McKerrow (1990).

The three reconstructions presented here include one each for the Devonian, Carboniferous, and Permian periods (Figs 6.2–6.4). The two most important events in the creation of Pangaea involved, in succession, the collision of Laurussia and Gondwana, and the collision of Siberia–Kazakhstania with Laurussia. The timing of the first collision has been a matter of some dispute. Geological evidence appears to favour the Late Devonian–Early Carboniferous time interval (Veevers

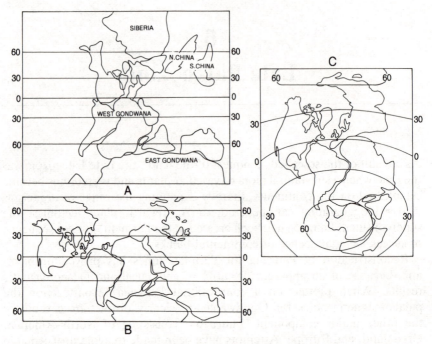

Fig. 6.1 Proposed Permian reconstructions of Pangaea. A, after Scotese *et al.* (1979); B, after Morel and Irving (1981); C, after Smith *et al.* (1981). After Hallam (1983b).

1988). This receives biogeographic support from Young (1990) who uses a land connection by Late Devonian times to account for the coexistence in Euramerica and eastern Gondwana of non-marine placoderm fish. Young explains that the close similarities between the Late Devonian fish of Morocco and eastern North America are inconsistent with the palaeomagnetically based model of Van der Voo (1988). This model involves the collision of Laurentia and Gondwana, including Avalonia, near the Silurian–Devonian boundary followed by separation of Gondwana from Laurentia (plus Avalonia) during the Devonian. Good faunal evidence from freshwater fish of initial terrestrial connections at or near the Frasnian–Famennian boundary, suggesting ocean closure, are in conflict with palaeomagnetic data suggesting an opening ocean. There is also the problem of explaining increasing faunal affinities between North America and Africa through Early and Mid Devonian times, the occurrence of the same genera in the Famennian of Ohio and Morocco, and the cause of the Acadian Orogeny.

The later collision, culminating in the Permian, involving Europe, Siberia, and Kazakhstania, to form the Urals suture was evidently complex,

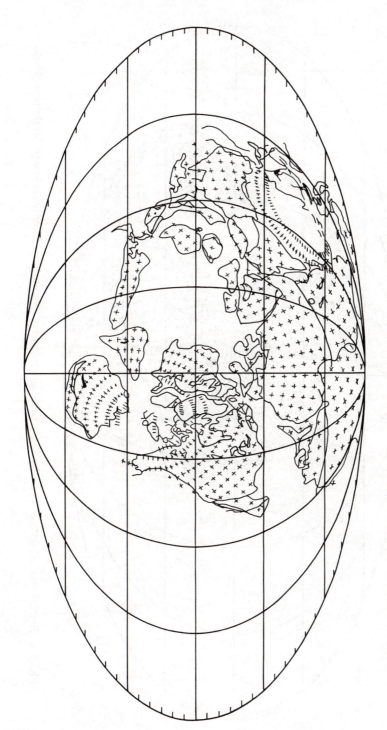

Fig. 6.2 Distribution of continents in the Mid Devonian (Givetian). After Scotese and McKerrow (1990).

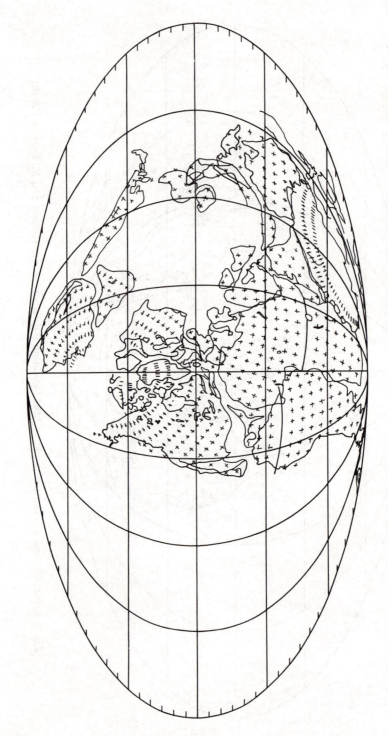

Fig. 6.3 Distribution of continents in the Early Late Carboniferous (Namurian). After Scotese and McKerrow (1990).

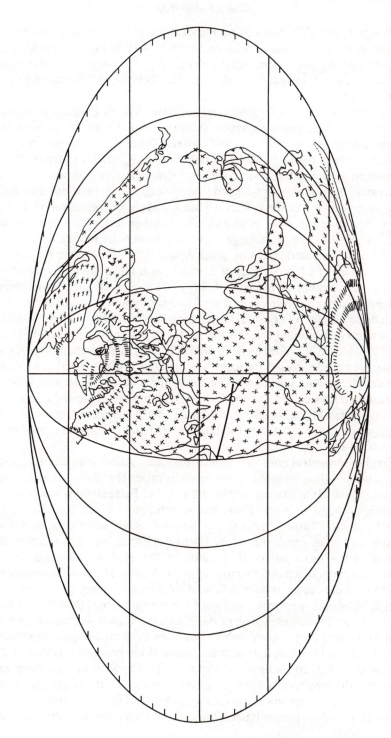

Fig. 6.4 Distribution of continents in the Late Permian (Kazanian). After Scotese and McKerrow (1990).

with significant differences in timing of the onset of collision in different places, over several tens of millions of years. A chronology of docking can be established from evidence of the shallowing of seaways, emergence, and massive influx of terrigenous sediments into depositional basins of (Talent *et al.* 1987).

The Scotese and McKerrow maps indicate that, as Pangaea coalesced, there was also a gross clockwise rotation, with Laurentia moving into progressively higher northern and eastern Gondwana into progressively higher southern latitudes. Older reconstructions ignored the plate tectonics of eastern Asia and the eastern limbs of Pangaea were portrayed as being separated by a major low-latitude ocean expanding in width eastward, accepted as some version of Edvard Suess's Mesozoic ocean called Tethys. Now that so much more is known about Asian tectonics it has become evident that this simple concept must be substantially modified.

The early reconstructions of Smith *et al.* (1973) showed the majority of Tibet and all of China as part of Eurasia, with the Zangbo Suture of the Himalayas marking the position of Tethys. However, all the known oceanic sediments along the Tethyan belt are Triassic or younger in age and no field evidence exists for a Permo–Triassic ocean (Sengör 1984). In Sengör's view a Cimmerian continent formed the northern margin of Gondwana in the Late Palaeozoic but rifted away in the Permian. Thus the site of the original Palaeozoic ocean, called Palaeotethys, lies north of the Kunlun orogenic belt and the North China block. It is the younger ocean, Neotethys, which is represented now by the Zangbo Suture. Others have suggested a sequence of small, strip-like terranes rather than a single continent Cimmeria. The Banggong Suture, separating the Lhasa and Qiantang terranes, could mark the site of Palaeotethys.

Precise reconstructions for Late Palaeozoic Asian continents are not yet possible. Biogeographic constraints support the theory that a series of microcontinents existed within the tropical Palaeotethys which subsequently accreted to Asia. These microcontinents include Sino-Korea (or North China), Tarim, Yangtze (or South China), Indochina, and East Qiantang, all of which shared the Cathaysia Flora. No united Cathaysian continent has been proposed, because of the existence of a number of tectonic sutures which did not close until much later. The general northward motion appears to have been followed by the rifting away of a number of peri-Gondwanan microcontinents in Late Permian times (Nie *et al.* 1990).

One possible reconstruction of the 'Cathaysian' microcontinents in relation to the rest of the world, which has been subjected to palaeomagnetic constraints and is consistent with the major floral provinces, is that of Lin and Watts (1988) and is shown as Fig. 6.5. The The Kunlun, Qiantang, and several Indochina terranes are grouped together with the North and South China blocks and all share the Cathaysia Flora. There is a wide spectrum of views on when particular terranes sutured together and their relative dispositions before closure. For instance did the North and South China blocks accrete as early as Devonian or as late as Triassic times (Smith 1988)? The relevant palaeobiogeographic evidence bearing on such questions will be considered later in this chapter.

According to current palaeogeographic reconstructions there was a shallow-water gulf in the western part of Tethys and the western end of Sengör's Palaeotethys is generally assumed to have been between the Cimmerian continent and stable Asia. However, evidence has recently been presented to indicate a belt of Permian deep ocean deposits extending from Sicily and Crete to Iraq, Oman, and the southern Pamirs; at least in Oman they directly overlie rocks indicating obducted oceanic crust (Catalano *et al.* 1991; Blendinger *et al.* 1992). The deposits contain conodonts, ostracods and radiolarians including species found in deep-water circum-Pacific deposits of the same age. This new information seems to imply that the Permian Tethys was situated adjacent to Gondwana to the south of the later Cimmerian continent, and had free communication with Panthalassa, the precursor of the present-day Pacific.

DEVONIAN

Because organisms confined to continents, or requiring land connections to allow migration, are often biogeographically more informative, non-marine vertebrates and plants will be considered before attention is turned to the more substantial record of marine organisms.

Non-marine vertebrates

Thelodonts are agnathan fish that had an armour of dermal denticles shed through life, as well as separating rapidly after death. In consequence articulated specimens are rare and, when found, clearly indicate a life assemblage. In the Early Devonian a new *Turinia–Apalolepis–Nikolivia* fauna appeared almost simultaneously in both continental and marine deposits of the northern hemisphere, and a very similar *Turinia* fauna also occurs in the Lower Devonian of Australia (Turner and Tarling 1982). The key question arises – was this group entirely freshwater or were marine phases involved? Articulated specimens are only found in freshwater or deltaic sediments so that it is indisputable that these thelodonts spent at least some of their lives in non-marine conditions, but whether or not this applies to their whole life cycle is uncertain. The apparent isolation of the Siberian thelodont fauna in the Early Devonian is of key importance in determining the life habit, because it implies that transoceanic or even coastal migration was unlikely. Since Turner and Tarling consider that

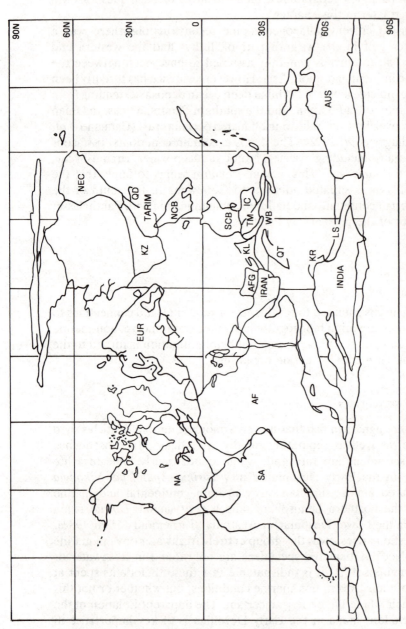

Fig. 6.5 A global reconstruction for the Carboniferous, based on palaeomagnetic results and guided also by the distribution of floras. AF, Africa; AFG, Afghanistan Terrane; AUS, Australia; EUR, Europe; IC, Indochina Terrane; KL, Kunlun Terrane; KR, Karakoram Terrane; KZ, Kazakhstan Terrane; LS, Lhasa Terrane; NA, North America; NCB, North China Block; NEC, North-east China Terrane; QD, Qaidam Terrane; QT, Qiangtang Terrane; SA, South America; SCB, South China Block; TM, Thai-Malay Terrane; WB, West Burma Terrane. Simplified from Lin and Watts (1988).

oceanic dispersal was not possible a land route is implied between Australia and Laurussia as early as the Early Devonian.

A more comprehensive review of Devonian vertebrate biogeography has been undertaken by Young (1981, 1990). Young considers that an assessment of which taxa are strictly freshwater as opposed to euryhaline can only be made through an analysis of distribution patterns in relation to phylogeny.

By the start of the Devonian four major groups of agnathans (A) and four major groups of gnathostomes (B) had become established. A, heterostracans, osteostracans, thelodontids, anaspids. B, osteichthyans (mainly crossopterygians and lungfish), chondrichthyans, acanthodians, placoderms.

Several provinces are distinguished for the Early Devonian (Fig. 6.6). *Euramerica* had a similar fauna characterized by osteostracans. The Nova Scotia fauna indicates that the Avalonia terrane was attached to Euramerica by Gedinnian times. This is consistent with Late Silurian ostracods from this region showing Baltic affinities but completely inconsistent with the suggestion by Van der Voo (1988) that the Acadian continent–continent collision was between Laurussia and Gondwana, with the Avalonian/Armorican terranes attached to the latter.

Siberia had a diverse and endemic fauna characterized by amphiaspid heterostracans. Palaeomagnetic data indicating higher latitudes, as in the Scotese reconstructions, are thought to be inconsistent with the presence of thick carbonate and evaporite sequences, signifying a tropical–subtropical latitude (Heckel and Witzke 1979). *Kazakhstan* is shown on the Scotese maps as an island continent between Siberia and North China but there is no evidence of the expected endemic fauna. The *South China* fauna is very distinctive, being characterized by diverse galeaspids and yunnanolepids. Apart from a few occurrences in North- China this fauna otherwise occurs only in north-eastern Vietnam, north of the Red River Suture. Such a highly diverse, presumed freshwater fish fauna is thought by Young to be best accounted for by an equatorial location.

Based on thelodonts and other groups it seems that the endemism of the *East Gondwana* region (based mainly on Australian evidence) persisted into the later Devonian (Givetian–Frasnian). Emsian marine placoderm faunas of eastern Australia show some affinities with South China. There was a probable continuity of faunas across Gondwana, and Bolivian faunas are distinct from those of Euramerica, implying the likelihood of a marine barrier, the so-called Theic Ocean. This palaeogeographic interpretation differs from that implied by Turner and Tarling.

In the Late Devonian regional provinciality declined. For example, whereas in the Early Devonian there had been strong isolation between South China and Australia, by the Late Devonian one endemic group of antiarch placoderms, the sinolepids, had reached eastern Australia.

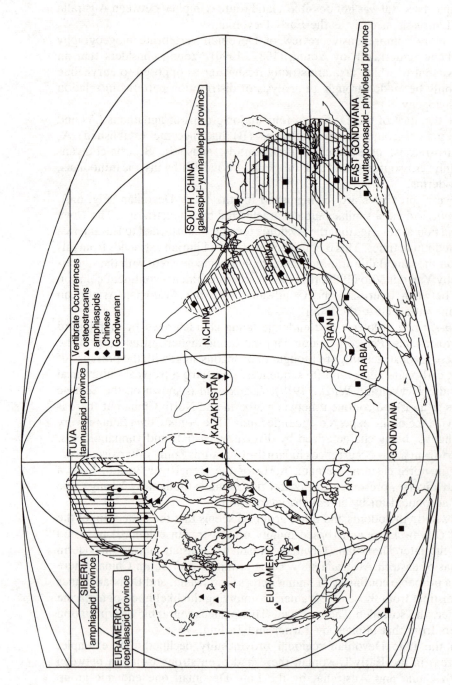

Fig. 6.6 Early Devonian vertebrate distribution patterns. Simplified from Young (1990).

Phyllolepid placoderms may have originated in East Gondwana. Their appearance in Euramerican strata at or near the Frasnian–Famennian boundary signifies the establishment for the first time of a persistent terrestrial connection. This group was absent from Siberia and North and South China, even though the other groups preserved were able to cross intervening barriers. For this reason phyllolepids have been interpreted as having been a strictly non-marine group (Young 1981). The Famennian vertebrates of East Greenland and Australia have at least six genera in common.

Terrestrial plants

The information derived so far from Devonian plants does not compare in quality with that from vertebrates. Edwards (1990) points out some of the difficulties in making adequate biogeographic studies. Information is for the most part derived from fragmentary allochthonous fossils and specialists are almost completely ignorant of the nature of whole plants. Furthermore, the simplicity of their axial organization is such that convergence must have been common, which provides a complication for taxonomic identification. There are also problems involving stratigraphic accuracy and precision in non-marine deposits, and of facies and the sorting of transported material. On the very limited data available for the Early Devonian, Edwards recognizes four regions that yield assemblages of distinctive composition—Laurussia, Siberia, Kazakhstan, and Australia.

Raymond (1987) has undertaken a more ambitious study of Early Devonian plants, using cluster and polar ordination analysis, her phytogeographic conclusions being based both on genera and morphological traits, which yield concordant results. Three major phytogeographic units are distinguished:

(1) equatorial–middle latitude, with South and North Laurussian and Chinese subunits;
(2) Australian;
(3) Kazakhstan–North Gondwanan.

Climatic differences are thought to account for the separation of Siberian–North Laurussian and Australian from Chinese and South Laurussian equatorial assemblages. Early Devonian marine evaporites in Siberia and Australia suggest a drier climate than Laurussia, which has no evaporites. Marine barriers to dispersal could have caused the separation of equatorial Chinese and Laurussian assemblages. The Kazakhstan–North Gondwanan unit poses a problem because most reconstructions show the two regions separated by ocean. Edwards (1990), however, rejects this unit as it was apparently based on a single misplotted assemblage.

Mid to Late Devonian (Givetian–Frasnian) phytogeography of Eur-america and western Gondwana has been studied by Streel *et al.* (1990) using miospore distributions. Their recognition of relatively uniform floras across a wide range of latitude is consistent with a relatively equable cli-mate, and close proximity is favoured between Euramerica and Gondwana. Like Young (1990) they challenge Van der Voo's (1988) claim of a newly opened ocean, over 2500 km wide, between Gondwana and Laurentia, with the Avalonian–Armorican terranes accreted to Laurentia. This is thought to be incompatible with the inferred pattern of spore dispersal.

Marine organisms

As in the Ordovician and Silurian articulate brachiopods are generally extremely abundant components of neritic faunas and their biogeography has been extensively studied. Johnson and Boucot (1973) recognize three provinces, the Old World, Appalachian, and Malvinokaffric, which were already in existence at the beginning of the period. The unhappily named Old World Province embraces western and Arctic North America, as well as Eurasia, North Africa, and Australia. It is characterized by Silurian holdovers such as eospiriferids, orthids, atrypids, and athyridids. In the Appalachian Province, Silurian holdovers are unimportant and a distinctive group of endemic taxa became established in the Gedinnian. As in the Silurian, the Malvinokaffric Province is marked by a very restricted fauna in which some important groups such as atrypids and gypidulids are unknown. Definitive genera include *Australospirifer* and *Australocoelia*. They are accompanied by typical Appalachian taxa, suggesting derivation from that province.

The greatest provinciality was in the Emsian (late Early Devonian). There are no Appalachian Province brachiopods in western North America but in Bolivia and Ghana there is a mixture of Malvinokaffric and Appalachian forms. During most of the Mid Devonian (Eifelian–Givetian) the contrast between the Appalachian and Old World Provinces is as strong as before, but in Late Givetian and Late Devonian (Frasnian–Famennian) times this provinciality disappeared. Use of Johnson's (1971) provinciality index gives a quantitative expression to the significant increase in cosmo-politanism in North America from the Mid to the Late Devonian. By Famennian times the Malvinokaffric Province had disappeared (no marine strata occur in either South America or Africa) and a globally cosmopolitan fauna was established.

As with most palaeobiogeographic studies, taxonomic comparisons of brachiopods have generally been confined to the generic level, but Talent *et al.* (1987) pay attention to species distributions in Asia and Australia and argue that there has been a more profound species provinciality through the Devonian and the Early Carboniferous than previously assumed. This is best explained, they believe, by the relative isolation of continental blocks. With regard to the Early Devonian, the modest but persistent level of similarity of species on the eastern and western slopes of the Urals is consistent with a model in which shelf faunas were separated by somewhere near a limiting 750–1000 km of deep ocean. They agree with Wang *et al.* (1984) about the distinctness of South China as a biogeographic region (based on brachiopods, trilobites, and rugose corals) but disagree with them when they state that the taxonomic distinctiveness declined in the Mid Devonian. This is because, according to Talent *et al.*'s analysis, the brachiopod species contrast was maintained until the Early Carboniferous, suggesting a profound isolation from the North China block.

A similar biogeographic pattern for North America is recognized for rugose corals (Oliver 1976; Oliver and Pedder 1979). The Lower and Middle Devonian corals of eastern and western North America, corresponding to Johnson and Boucot's Appalachian and Old World Provinces, are markedly different, with a maximum endemism, involving a maximum of 91 per cent of genera, in the Late Emsian. The degree of endemism subsequently diminished through the Mid Devonian to zero in the Frasnian. Pedder and Oliver (1990) subsequently reviewed global distributions of rugose corals. They find, in general, good agreement with an earlier Emsian reconstruction of Scotese but note two anomalies. The large but poorly known faunas of Mongolia and the Amur Basin occur at approximately 60°N and the well known faunas of Altai-Sayan at 45–50°N in this reconstruction. In the light of the known distribution of both modern corals and Devonian southern hemisphere corals Pedder and Oliver consider it questionable that the original latitude of any large Devonian northern hemisphere fauna would have been greater than 45°. Perhaps in response to this criticism, the new Emsian reconstruction of Scotese and McKerrow (1990) shows Mongolia close to 45°N.

Other neritic groups that show pronounced Early Devonian endemism include trilobites, echinoderms, and nautiloids. As noted in the previous chapter, the Malvinokaffric Province was originally named on the basis of endemic trilobites. According to Eldredge and Ormiston (1979) 80 per cent of the Lower Devonian trilobites of Bolivia are endemic at subgeneric level; they consist mostly of calmoniids and dalmanitids. The pattern of provincialism of Early Devonian echinoderms resembles that of brachiopods; by Mid Devonian times provincialism was starting to decline (Witzke *et al.* 1979). Crick (1990) recognizes three nautiloid provinces in the Gedinnian. Orthocerids dominated in Laurentia, oncocerids in Baltica, and discocerids in Siberia and Kazakhstan. By Frasnian times the Kazakhstan fauna became significantly similar to that of Baltica, for the first time, and the fauna of Western Australia more closely resembled that of central Europe and Turkey.

Among more mobile organisms endemism is less marked and distinct

provinces are recognizable neither in goniatites (House 1973) nor conodonts (Klapper and Johnson 1980). In both cases such endemism as there
was declined later in the period, as with the other groups reviewed above.
It is noteworthy in this context that eastern North American goniatite
faunas are restricted in diversity, especially in the classic deposits of New
York State, with many common European genera not known, though a
number occur in the Cordilleran region of North America. Further south
in the Appalachians more taxa occur, such as *Sobolewia* and *Mainioceras*,
strongly suggesting a direct link with North Africa. Another strong link
is inferred between the Western Cordillera of Canada and the Urals
and Novaya Zemlya, indicating a free marine seaway on the northern
side of Laurussia. The Devonian faunas of Western Australia show very
strong European affinities, indicative of a proto-Tethys link. The most
cosmopolitan distribution of goniatites was in the Frasnian.

The endemism and dispersal of Devonian conodonts received a comprehensive treatment from Klapper and Johnson. The distribution of
conodonts was more or less pantropical and none are found in the
Malvinokaffric Realm. Most conodont genera are cosmopolitan but endemism is shown at the species level. This is a clear difference from
such groups as corals and brachiopods. Certain species, especially of
Icriodus, characterized shallow, nearshore environments while others,
for example of *Polygnathus* and *Palmatolepis*, lived in deeper, more
offshore areas. Conodonts indigenous to nearshore biofacies tend to have
been isolated in separate epicontinental seas whereas those in offshore
biofacies were more widely distributed or fully cosmopolitan. To account
for distributional patterns Klapper and Johnson invoke a south equatorial
marine current along the southern margin of Euramerica (or Laurussia),
which would also help to account for the common occurrence of Early
Devonian trilobite and brachiopod species in Europe and eastern North
America; this could have been along an epicontinental seaway. A north
equatorial current on the northern side of Euramerica is also invoked,
to account for the inferred free communication between cordilleran and
Arctic North America, various Asian microcontinents and Australia.
Endemism declined significantly from the Early to the Late Devonian,
as with other groups.

Flessa and Hardy (1988) have undertaken a cluster analysis of Klapper
and Johnson's data, using Jaccard similarity coefficients. They establish that
provinciality was initially moderate (in the early Lochkovian), declined to
low levels in the Late Lochkovian to Early Pragian, and increased to a
maximum in the Emsian and Eifelian. Post-Eifelian faunas show little
biogeographic differentiation. Provinciality, when developed, is usually
expressed by differentiation into western (north-west Laurussia) and
eastern (proto-Tethys) regions. Australian faunas, when present, are
usually distinct.

Causes of the changing provinciality through the period

It is clear from virtually all the marine groups studied that provinciality increased significantly in the Early Devonian from a minimum amount in the Silurian and thereafter decreased later in the period. Cocks and McKerrow (1973) attributed the Early Devonian provinciality increase to an increase in climatic zonation, but this proposal receives no support from independent palaeoclimatic evidence (Frakes *et al.* 1992). Oliver (1976) more plausibly relates the waxing of endemism in the eastern Americas to the build up of a land barrier on the south-eastern side of the Appalachian belt, due to the collision of Laurussia and Gondwana.

More generally, changing relative sea level appears to be the most likely cause of increasing and then declining provinciality in the Devonian. The Early Devonian marks the time of one of the most significant sea-level low stands in the Palaeozoic (Fig. 4.1). As sea level rose from an Emsian minimum (and provinciality maximum) to a Late Frasnian maximum (and provinciality minimum) intervening land barriers were progressively flooded and interconnections established between hitherto relatively isolated epicontinental seas (Johnson *et al.* 1985). This is explicitly stated in the model of Klapper and Johnson (1980).

Interestingly, regional endemism declined through the course of the Devonian for non-marine vertebrates (Young 1990). This cannot be attributed to the same factor as for the marine groups, and is more likely to be a response to the coalescence of continents.

Copper (1986) attributed the mass extinction at the Frasnian–Famennian boundary, involving among other things the destruction of coral–stromatoporoid reef ecosystems, to a global temperature fall, and cited in support the occurrence of Famennian glacial deposits in Brazil (Caputo and Crowell 1985). This poses a problem, because the stronger climatic zonation in the Late Devonian that would be a natural consequence of this should have given rise to increased rather than decreased endemism. In particular, neither the occurrence of *Bothriolepis* over a wide range of palaeolatitude, from 60°N to 65°S in the Scotese reconstruction (Young 1990), nor the occurrence of relatively uniform floras throughout Gondwana and Euramerica (Streel *et al.* 1990) are compatible with Copper's cold-climate model. It is possible, indeed, that the Brazilian glacial deposits have been misdated, being actually Carboniferous in age (M.R. House, personal communication).

CARBONIFEROUS AND PERMIAN

A significant change in global climate occurred during the Carboniferous and Permian periods, with extensive Gondwana ice sheets commencing their growth in the Mid Carboniferous and disappearing in the Mid Permian (Frakes *et al.* 1992 and Fig. 4.1).

Ross and Ross (1985) see a direct relationship between changing world geography and the onset of cooler climates. By the Carboniferous, Gondwana and Euramerica had joined to form a north–south orientated supercontinent and in the Early Permian Siberia coalesced with Euramerica to form Pangaea. These changes are thought by Ross and Ross to have disrupted circum-equatorial warm temperature tropical seas to develop a north–south directed surface-water circulation accompanied by cooler world climates. Dispersal of tropical and subtropical benthic invertebrates became increasingly difficult and intermittent as the Franklinian Shelf of northern Euramerica moved progressively northwards into cooler waters. The formation of Pangaea curtailed further dispersal of tropical benthos around the northern margin of this supercontinent.

A different view is presented by Ziegler *et al.* (1981). According to them the deflection of equatorial currents to the north and south caused tropical oceanic conditions to extend throughout Tethys. They claim that there was an increasing provinciality from the Early Carboniferous to the Permian, probably related to increased climatic gradients and the elaboration of tectonically produced barriers to dispersal. However, as Bambach (1990) has pointed out, marine diversity declined during this time, whereas it should have increased significantly by the increased provinciality induced by stronger climatic gradients (Valentine 1973; Valentine *et al.* 1978).

Bambach undertook an extensive analysis of a large number of invertebrate groups and concluded that there was in fact no marked change in endemism of the total marine fauna from the Early Carboniferous to the Early Permian. The major change was the loss of distinction between the tropical European and Chinese realms of the Carboniferous and the emergence of a Tethyan Realm in the Permian. A high-diversity tropical zone persisted from the Early to the Late Carboniferous but an intermediate diversity zone disappeared and low-diversity faunas characterized all latitudes greater than 15° south and 20° north. The South-east Asia block increased markedly in diversity in the Permian as it moved into the tropics.

Terrestrial plants

The record of terrestrial plants is significantly greater than for the Devonian and is highly informative as to provinciality. Most attention has justifiably been paid to Late Carboniferous and Permian floras but before these are dealt with attention should be paid to Raymond's (1985) study of Early Carboniferous floras in the northern hemisphere (Gondwana assemblages were excluded from her analysis because they are poorly dated). Floras of three time intervals were studied:

(1) Tournaisian–Early Visean
(2) Visean
(3) Late Visean–earliest Namurian A.

Raymond demonstrates a loss of phytogeographic differentiation within equatorial to middle latitudes between intervals 2 and 3. Increases in Siberian diversity and northward expansion of palaeolatitude limits in the range of equatorial genera suggest that Siberia experienced climatic amelioration in unit 3. A predicted effect of Laurussia–Gondwana collision is climatic amelioration in northern middle and high latitudes. Raymond argues therefore for collision in the Late Visean, a time considerably later than other researchers such as Young (1990) demand. The onset of southern hemisphere glaciation in the Early Namurian does not contradict this hypothesis of climatic amelioration in the region bordering Tethys, because the Early Carboniferous climatic belts were probably asymmetric, with a relatively warm, ice-free north pole and cold, continental south pole.

It is clear that, while global climate was relatively uniform in the Early Carboniferous (Raymond 1985) it had become highly differentiated by the latest Carboniferous and Permian, when the regionalization of floras was more pronounced than at any earlier phase of land plant evolution (Chaloner and Creber 1988). Four provinces are distinguished (Chaloner and Meyen 1973; Chaloner and Lacey 1973; Plumstead 1973; Fig. 6.7):

(1) Euramerican
(2) Cathaysian, or *Gigantopteris*
(3) Angara
(4) Gondwana, or *Glossopteris*.

As early as 1937, Halle interpreted the *Glossopteris* flora of India as being the result of post-Permian continental drift that brought it into juxtaposition with the Cathaysian flora. He also recognized the difference between the Angara and Cathaysian floras and their relationship to the Euramerican flora.

Chaloner and Creber (1988) emphasize the problems of studying fossil plants from a biogeographic standpoint, such as the separation of organs, transportation from their growth site and different preservation states. Moreover, it is dangerous to utilize leaves alone. The leaf genus *Phyllotheca* has been widely reported from both the Angara and Gondwana provinces but it has been shown that, despite the similarity of leaves, the fructifications (spore-bearing structures) are of fundamentally different types of Equisetales. Furthermore, *Glossopteris* has a distinctive but not unique venation. There has been a Russian report of *Glossopteris* from Siberia but without supporting evidence of fructifications the relationship of these Angaran leaves to the Gondwana *Glossopteris* remains problematic.

There is a continuity of the Cathaysia flora from South China through

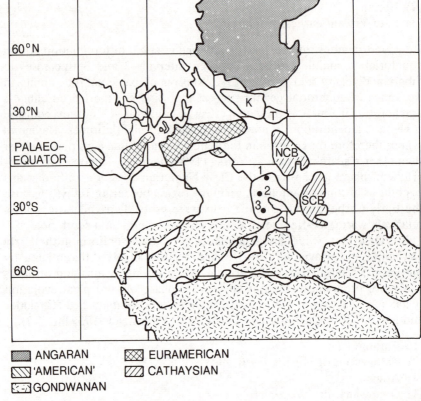

Fig. 6.7 Permian reconstruction of plate positions, showing distribution of floral provinces. K, Kazakhstan block; T, Tarim block; NCB, SCB, North and South China blocks. 1, Hazro, Turkey; 2, Ga'ara, Iraq; 3, Unaizah, Saudi Arabia. These floras have some Cathaysian affinity. After Chaloner and Creber (1988).

Thailand and Malaysia into Sumatra, which implies continuity of land. Western New Guinea has *Glossopteris*, implying that this region must have been juxtaposed to Australia. This is scarcely compatible with the Carboniferous–Triassic reconstructions of Audley-Charles (1983) which showed a sustained juxtaposition of Sumatra and New Guinea. Chaloner and Creber prefer the reconstruction of Lin *et al.* (1985) which shows western New Guinea on the northern edge of the Australian plate and the South China block a long way to the north in tropical latitudes (Fig. 6.7). The *Glossopteris* flora also occurs in southern Tibet, so the Cathaysia–Gondwana junction must lie within Tibet.

Most if not all the genera characteristic of the Cathaysia Province occur on both the North and South China blocks and the differences between these regions are of relative frequency. Thus *Gigantopteris* is widespread in the south but rare in the north, while the reverse is true of *Taeniopteris*. Chaloner and Creber consider that, despite differences at specific level, regional differentiation is modest, but consistent with their being on separate plates, though the floral evidence hardly requires it. Laveine *et al.* (1987) note that numerous floral taxa are common to both North and South China blocks, throughout the Carboniferous and Permian, with relatively few taxa being geographically confined. They therefore support the claim by Mattauer *et al.* (1985) for a relatively early collision, in the Devonian.

By applying the modern plant biome concept Ziegler (1990) has attempted a comparison between Permian and Quaternary climatic zones, distinguishing ten biomes. The Lycophyta (club mosses) and Sphenophyta (horse tails) are extant as herbaceous forms. Arborescent forms of both groups were important elements in tropical and warm temperate environments. Tropical lycophytes seem to have been coal-swamp dwellers. Sphenophytes were much more widespread and diverse. Most of them were herbaceous; arborescent forms are limited to the tropics; as none have been found in the Gondwana or Angara floras. Filicales and Marattiales were well represented, many of them tree ferns. Pteridospermaphyta (seed ferns) are all extinct today. They were mainly restricted to warmer climates, with gigantopterids being confined to tropical rainforests. Glossopterids, usually classified with pteridosperms, and cordaitids are confined respectively to the high latitude Gondwana and Angara Provinces. They may exhibit parallel evolution as a consequence of adaptation to cold climates, with features such as tree trunks with dense wood and simple, tongue-shaped, probably deciduous leaves. The Voltziales, transitional to modern conifers, had reduced needle-like to scaly leaves, evidently adaptations to drought, because they are often found in regions transitional to evaporite basins.

The term Euramerican Realm or Province has been applied to tropical rainforests of the Carboniferous and to floras peripheral to rainforests in the Permian. Euramerica moved out of the equatorial zone at about the Carboniferous–Permian boundary and rainforests became restricted to Cathaysia. Ziegler distinguishes the following realms (Fig. 6.8).

Tropical or Cathaysian—biomes 1, 2, 3. Biome 1, the Gigantopteris flora, is characterized by high generic and ordinal diversity, a common association with coal and equatorial palaeolatitudes. Besides occurring in North and South China, Indochina, Tarim, and Qiantang there are outliers in Arabia, Spain, Venezuela, and Texas. Genera and even species are widespread between China and North America and hence it is believed that the Asian continents were geographically proximal to Pangaea. 'Summerwet' biome 2 is marked by the Euramerican, Zechstein, and *Callipteris* flora and is dominated by Voltziales. Biome 3 is desert, characterized by evaporite deposits.

North Temperate and Angaran—biomes 4, 5, 6, 7, 8. The cool temperate biome 6, the Angara flora, is restricted to Siberia and Mongolia, which were therefore contiguous in Permian times. The ordinal diversity is restricted but the generic diversity of cordaitids and sphenophytes is high. This biome is based on the assumption that the cordaitids were deciduous and the sphenophytes herbaceous. Biome 7 is mid latitude desert and 5 warm temperate. The junction of 5 and 6 is probably coincident with the line marking the incidence of severe frost.

South Temperate or Angaran—biomes 5, 6, 7, 8, 9, 0 (equivalent to 10—glacial). Glossopterids, interpreted as deciduous trees, and temperate *Glossopteris* flora. Little else besides *Gangamopteris* is assigned to the cold temperate biome 8.

Non-marine tetrapods

By comparison with the plants there has so far been very little biogeographic study of vertebrates. Milner and Panchen (1973) carried out a comparison of the Permo–Carboniferous faunas of Europe and North America. The terrestrial faunas are very similar but the aquatic faunas show substantial differences in the latest Carboniferous (Stephanian) and Permian, with different families filling the same ecological niches in the two regions. The same faunal divergence was also present on a smaller scale in the Late Westphalian, at which time much of the aquatic fauna was still geographically uniform. This suggests a partial barrier to aquatic, but not terrestrial, tetrapods, manifested more extensively in the Stephanian and Permian. The most likely barrier would have been a range of uplands such as produced by the Hercynian Orogeny.

Marine organisms

By Late Carboniferous and, more especially, Permian times a low-latitude Tethyan Realm had become clearly distinct from higher latitude biota. It was characterized by higher diversity of all the organisms that have been comprehensively analysed from a biogeographic point of view, including brachiopods (Waterhouse and Bonham-Carter 1975), bryozoans (Ross 1978), foraminifers (Yancey 1979; Okimura *et al.* 1985; Ishii *et al.* 1985) and conodonts (Matsuda 1985). Corals and fusuline foraminifers are confined to the Tethyan realm, and calcareous smaller foraminifers dominate over

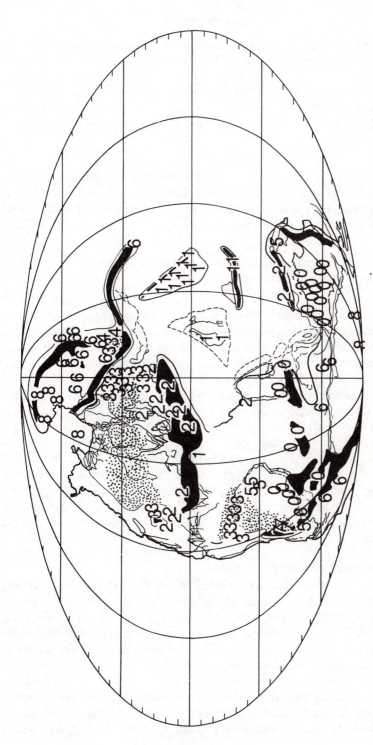

Fig. 6.8 Early Permian (Sakmarian) biome distribution. The biomes are represented by the digits 1 to 9 and 0, with 0 representing the number 10. The black areas represent mountains, uplands, and glaciers. The dot pattern represents desert areas, with the digits 3 and 7 representing the desert biomes; the placement of the digit 0 is based on ice sheet margins. Full explanation in text. Simplified from Ziegler (1990).

arenaceous forms, whereas the reverse is true in the inferred temperate or cooler regions (Okimura *et al.* 1985).

The most thorough quantitative study that illuminates the change in biota with palaeolatitude is that of Waterhouse and Bonham-Carter (1975) who undertook a Q-mode cluster analysis of the global distribution of Permian brachiopod families. They distinguished three groups, noting associated faunas and facies.

Group A is characterized chiefly by martiniids, with linoproductids and spiriferids also rating strongly. There are virtually no corals or fusulines and there are often associated glacial deposits. Eurydesmatids and deltopectinids are characteristic bivalves.

Group B is characterized chiefly by marginiferids and echinoconchids; some corals and fusulines occur.

Group C has the highest diversity and has many associated corals and fusulinids. The most diagnostic families are the meekellids, aulostegids and two highly modified ones with coralloid form and vesiculate shell spaces, the richtofenids, and lyttonids. The sedimentary facies has abundant carbonates, unlike Groups A and B.

Palaeogeographically, the 'tropical belt' with persistent Group C features extends through the Mediterranean region, Crimea and Greater Caucasus, the Pamirs, South China, Indochina, and Texas. The 'temperate belt' of Group B is very widespread, embracing much of western North America, northern Canada, northern Russia, Spitsbergen, and northern Europe, together with Bolivia, Brazil, and parts of the Himalayas. The 'cold temperate' belt (Groups A and B) includes Greenland, the Taimyr Peninsula of northern Siberia and New Zealand. Finally, the 'polar' region, characterized by Group A, is represented by eastern Australia.

Variations through time and space.

Based on an analysis of bryozoan distributions Ross and Ross (1990) infer some significant changes through the course of the Carboniferous and Permian. In Early Carboniferous (Tournaisian and Visean) times the climate was relatively warm, sea level comparatively high and bryozoans commonly had widespread distributions. Late in this subperiod Gondwana and Euramerica are considered to have joined along the Hercynian–Appalachian–Marathon orogenic belt to form Lesser Pangaea. This is inferred to have resulted in the elimination of the tropical marine shelf along the southern margin of Euramerica and the marine connection that linked the neritic faunas of western Euramerica with the eastern proto-Tethys. In consequence two tropical shelf faunas became isolated, one on either side of Lesser Pangaea, and dispersals between them were infrequent along a warm temperate route on the northern Franklin shelf of Euramerica.

In mid Early Permian times, Lesser Pangaea joined with Angara (= Siberia) along the Urals, so that tropical shelf faunas on either side of Greater Pangaea became effectively isolated. Each fauna evolved independently, with only occasional dispersal by island hopping across Panthalassa. This explains the strongly provincial faunas of the later Permian, enhanced in Guadaloupian times by increased diversity.

Ross and Ross's interesting model needs to be tested by reference to other groups. One problem concerns the precise timing of the creation of 'Lesser Pangaea' by the collision of Gondwana and Euramerica (or Laurussia). As has been noted earlier, estimates have ranged between Late Devonian and late Early Carboniferous (or even, more tentatively, Early Devonian if Turner and Tarling's (1982) inference is correct). The problem may be resolvable by reference to the distinction between true ocean and epicontinental seaway. The latter would permit migration of neritic faunas but would be intermittent, disappearing at times of regional tectonic uplift and/or sea-level fall. Continental collision could have been a Devonian event, and the significant east-west restriction of tropical fauna migration could have been the consequence of tectonic or tectonoeustatic events towards the end of the Early Carboniferous, probably associated with the inception of the Gondwana ice sheets (Veevers and Powell 1987).

Within the Tethyan Realm of Permian times Yancey (1979) makes a distinction between an Old World Tethyan Province with waagenophyllid corals and verbeekinid fusulines, and a Grandian Province (Texas and New Mexico) lacking these taxa. Particular interest attaches to the East Asian region in current palaeobiogeographic and tectonic research. In Yancey's opinion the most conspicuous feature for the Permian is the lack of a temperate province, with no gradual trend from tropical to polar biotas as can be recognized in Europe and North America. This provides important evidence for a major north–south continental closure of a minimum of 25°, with North China having moved northward from the tropical zone. The boundary between the Boreal and Tethyan Realms in East Asia is in effect a narrow zone near the northern border of China extending to northern Honshu (Japan). Ishii *et al.* (1985) provide more detail on the distribution of Permian fusulines. Whereas *Neoschwagerina* is widely distributed throughout the Tethyan Realm *Colania* and *Lepidolina* were restricted to the present East Asian region.

Nakamura *et al.* (1985) distinguish three brachiopod provinces in East Asia, which became clearly differentiated in Mid Permian times.

1. Angara Tethyan Province, occupying the southern border of the Angara continent, characterized by *Yakovlevia* and *Horridonia*. The North China fauna has resemblances to the South China fauna in the presence of such genera as *Leptodus* and *Richtofenia*.

2. Middle Tethyan Province (Cathaysia Tethyan Subprovince) of South China and Indochina, with *Monticulifera, Permianella*, and other characteristic genera.

3. Gondwana Tethyan Province, of the Himalayas and Salt Range of Pakistan, with *Costifera*.

In the Late Permian the Angaran Tethyan Province was lost because of widespread emergence. The strata in Kashmir and the Salt Range contain many more Tethyan elements than in the Mid Permian, such as *Oldhamina* and *Orthotetina*, in association with dominant Gondwana ones, but the region south of the Zangbo Suture undoubtedly belonged to the Gondwana Tethyan Province throughout the Permian. The genera *Peltichia* and *Cathaysia* are endemic to South China, suggesting a continued degree of isolation even near the end of the Palaeozoic.

The most rigorous biogeographic analysis of East Asian Carboniferous and Permian marine faunas is that of Smith (1988). Accordingly it warrants detailed attention. Smith made a study of rugose coral genera using three different quantitative techniques which gave broadly concordant results, parsimony analysis of endemicity, principal co-ordinates analysis, and single linkage cluster analysis. Data were studied for four time intervals.

1. *Late Early Carboniferous (Visean).* A broad central band of high diversity extended from South-east China through the Quilian and Kunlun Terranes to Kazakhstan and the southern Urals. The diversity was less on either side (North-eastern China and Qiantang Terranes) and the lowest diversity was in the Lhasa and Himalayas to the south and eastern Siberia to the north. Compound corals are the major components in the high diversity faunas but more or less absent from the very low diversity faunas. The most significant feature is the marked difference between the fauna of Kazakhstan and the Urals and Qilian–Kunlun–South-east China. This implies some form of barrier, with the Tienshan orogenic system marking the approximate boundary.

2. *Early Early Permian (Asselian/Sakmarian).* This was a time of extensive glaciation in the southern hemisphere and unsurprisingly coral faunas are few and diversity low, with a pattern of distribution similar to the Visean. The occurrence of only simple solitary corals in relatively shallow-water deposits of the Qiangtang, Lhasa, and Himalayan Terranes suggests that they lay outside the equatorial zone which, compared

with the Visean, was less extensive. This accords with the evidence of glaciation.

3. Late Early Permian.   Most Rugosa are simple solitary corals, with the faunas of the Urals differing significantly from those of 'Cathaysia'. The appearance of fusulines in the Lhasa Terrane may be the first indication of disjunction between this and the Himalayan Terrane.

4. Early Late Permian.   The highest diversities are in South-east China and the Lhasa Terrane, where compound and complex solitary corals predominate; these are the most similar faunas. There was a sharp faunal discontinuity between the Lhasa and Himalayan Terranes, indicating a barrier to dispersal.

Some general conclusions are drawn by Smith from his analysis and a review of other data. All of 'Cathaysia' (well defined by the flora) south of the Tienshan–Hegen Suture and north of the Banggong Suture, formed a distinctive equatorial region at least until Mid Permian times. The barrier between the Cathaysian and Angaran flora broke down at the end of the Permian. No such mixed affinity Gondwana–Cathaysian floras are known in the south.

The most significant feature of the faunal and floral distributions is the marked distinction between broadly tropical biotas in the Lhasa, Qiangtang, Kunlun, and North and South China Terranes compared with the Himalayan Terrane in the south and Mongolia–eastern Siberia in the north. The southern boundary was gradational in the Visean and it does not coincide with a single suture zone through time. The fact that the Lhasa Terrane may at one time have had a 'cold water'-type fauna comparable to the Himalayas and, shortly afterwards, have been similar to Cathaysia suggests that the primary control was climatic. This is supported by the fact that the Qiantang Terrane straddled the boundary between Himalayan and Cathaysian biotas in the Early Permian. Thus no suture can be taken as the site of a major ocean barrier.

Smith interprets the whole region as a northern equatorial extension of Gondwana and does not favour a series of scattered islands as portrayed in the Lin and Watts reconstruction (Fig. 6.5) because of the lack of deep ocean sediments associated with any of the proposed island terranes. The palaeomagnetic data on which this island reconstruction is based indicates a discrepancy of about 6° latitude between the North and South China blocks at the end of the Permian, but this could easily have postdated Early to Mid Permian breakup of the region.

Sengör (1984) placed the northern margin of his Cimmerian continent along Kunlun and the northern boundary of North China. However, having the Tienshan–Yinshan Suture as the site of Palaeotethys would explain why this was a biogeographic barrier for corals and plants in the Late Palaeozoic. The Cimmerian block would have been a northward continuation of Gondwana into the tropics, with a relatively small Palaeotethys forming a broadly equatorial barrier separating northern and southern tropical/subtropical zones.

7

Early Mesozoic

In this book a distinction has been made between Early Mesozoic (Triassic and Jurassic) and Late Mesozoic (Cretaceous) because of the very different global geographies of these times. As already indicated, Pangaea had become established as a coherent supercontinent at the beginning of the era and did not disintegrate significantly until the Cretaceous.

Although uncertainties remain, especially in Asia, knowledge of the relative positions of continental areas in the Mesozoic is considerably better than for the Palaeozoic. A prime reason for this is the existence for the first time, from the Mid Jurassic onwards, of a deep sea record, with magnetic anomalies, preserved on the present ocean floor. Reconstructions have been made starting from the present and removing successively older areas of oceanic crust. This is done by matching sets of magnetic anomalies and following traces of fracture zones on the ocean floor. As for the Palaeozoic, palaeomagnetic data from the continents provide latitudinal control. The early set of reconstructions of Smith and Briden (1977) are still valuable as first-order approximations of reasonable accuracy (Figs 7.1 and 7.2) except for Asia and parts of the Pacific margins, where a series of terranes was accreted. Furthermore, the continental fits are unduly tight, having failed to allow for the fact that there can be significant continental extension before breakup.

Even within the western part of Pangaea there are two areas where the fit has been awkward and controversial. The well-known problem of overlap between South and Central America in the original Bullard *et al.* (1965) reconstruction can be reconciled by restoring large displacements on a major north-west–south-east left-lateral fault in Mexico (Coney 1979). The reconstruction of Gondwana based on analysis of geophysical data (Norton and Sclater 1979) also runs into overlap problems, with the Antarctic Peninsula superimposed on the Falkland Plateau, which is known to be composed of continental crust. However, neither geology nor palaeomagnetism support the alternative proposed by Harrison *et al.* (1979), with the Antarctic Peninsula lying on the Pacific side of southernmost South America (Dalziel and Elliott 1982). Dalziel and Elliott believe that resolution of the problem is best achieved by recognizing that West Antarctica is probably a continental mosaic composed of several former microplates. Relative motion involving both translation

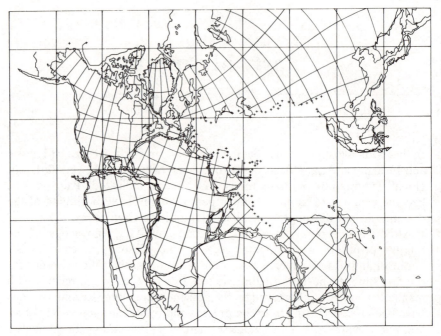

Fig. 7.1 Reconstruction of Pangaea in the Late Triassic. After Smith and Briden (1977).

and rotation probably took place during Gondwana breakup. This can explain the apparent overlap with the Falkland Plateau.

In the ensuing account, the topics of Pangaea breakup and Asian accretion will be addressed first. Only after a biogeographic review will the other important geological topics of circum-Pacific displaced terranes and marine corridors across western Pangaea be addressed.

INITIATION OF BREAKUP OF PANGAEA

In the Late Triassic and Early Jurassic a widespread belt of doming and rifting divided Laurasia and Gondwana along a line extending from the present Gulf of Mexico and Central Atlantic to the western part of Tethys. The first phase of creation of oceanic crust did not take place until early in Mid Jurassic time (Savostin *et al.* 1986; Ziegler 1987). The oldest known sediments directly overlying oceanic basalts are Callovian in age, from the Blake–Bahama Basin north-east of Florida. North-west Africa pulled away from eastern North America, with associated transcurrent fault movement, to create a narrow, initially rather isolated ocean basin in the Central Atlantic (Fig. 7.2). Rifting in the young Atlantic propagated south-westwards and opened a deep-water passage into the Pacific in the Late Jurassic, a passage which continued to widen in the Cretaceous (Klitgard and Schouten 1986 and Fig. 7.13).

The precise timing of opening of an American seaway from the western Tethys to the Pacific has been a controversial matter (Jansa 1991). Estimates have ranged from as early as Pliensbachian to as late as Callovian. The opening changed the hydrology of the Central Atlantic from a semi-restricted gulf into an open channel system, but this could have been achieved by an epicontinental seaway as well as by a truly oceanic strait. Biogeographic data have an important role to play in resolving this geological question, as will be demonstrated later.

The other important event in the early phase of Pangaea breakup was the initiation of separation of the eastern and western components of Gondwana, early in the Late Jurassic (Fig. 7.3). This dating is based primarily on a series of ocean-floor magnetic anomalies discovered in the Mozambique Basin (Veevers *et al.* 1980). Such separation would have allowed the establishment of a direct oceanic link between Tethys and the Pacific, between South America and Antarctica.

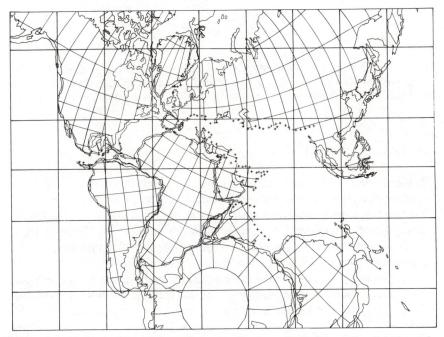

Fig. 7.2 Reconstruction of Pangaea in the Late Jurassic. After Smith and Briden (1977).

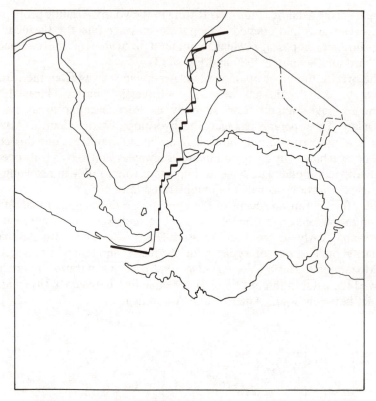

Fig. 7.3 Gondwana reconstruction showing line of opening at about 150 Ma. Ridge segments shown by thick line, transform faults by thin line. Dot–dash and full line show the minimum and likely northward extent, respectively, of Greater India. After Veevers *et al.* (1980).

ASIAN ACCRETION

A consensus has emerged in recent years that much of South-east Asia consists of terranes rifted from the margin of eastern Gondwana (Fig. 7.4) but there has been disagreement about the timing of rifting and collision events (Audley-Charles 1983; Audley-Charles *et al.* 1988; Metcalfe 1988, 1991; Sengör, 1984; Sengör *et al.* 1988). Sengör's view is that an elongate continental strip—Cimmeria—rifted away from the Gondwanan margin in Late Permian times and accreted to Eurasia by the Mid Jurassic. Suturing of Indochina to South China, and of Sibumasu (Burma, north-west Thailand, West Malaysia, north-west Sumatra) to Indochina was accomplished by the Late Triassic. Audley-Charles favours a Late Carboniferous to Early Permian rifting of Indochina and central Tibet (attached to Turkey and Iran) from Gondwana. It was followed after a considerable time interval by rifting of other South-east Asian continental blocks together with South Tibet in Mid to Late Jurassic times. Metcalfe favours an Early Triassic suturing of Sibumasu to 'Cathaysialand' followed by collision with North China in latest Triassic to Early Jurassic time, an interpretation apparently supported by palaeomagnetic data.

Extensive Late Carboniferous to Early Permian diamictites in Sibumasu and the Lhasa Block of South Tibet are most plausibly interpreted as glaciomarine sediments, and there are many strong biogeographic affinities supporting a peri-Gondwana location in the Late Palaeozoic. On the other hand both Palaeozoic and Mesozoic faunas and flora of Indochina, embracing East Thailand, Laos, Kampuchea, and most of Vietnam, are of warm climate type with strong affinities to North and South China (e.g. the Late Palaeozoic Cathaysia Flora). Triassic floras have affinities with Yunnan (Kimura 1984) and both Upper Triassic phytosaurs and lungfishes and Lower Jurassic crocodiles indicate Laurasian affinities (Buffetaut 1985a); a land connection must have existed.

Metcalfe (1988) disputes Audley-Charles' (1983) view about later Jurassic suturing of Sibumasu and associated continental blocks. He points out that

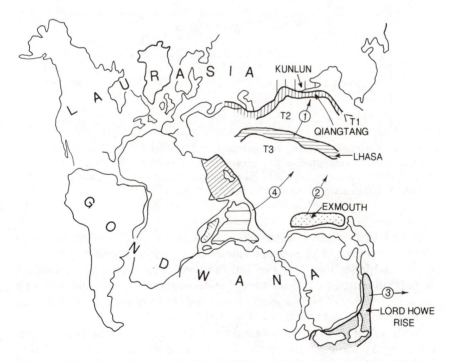

Fig. 7.4 Palaeotectonic reconstruction of Tethys for the Early Jurassic, illustrating the various strips and blocks of continental lithosphere (ornamented) that detached from the Gondwana margin. After Dewey (1988).

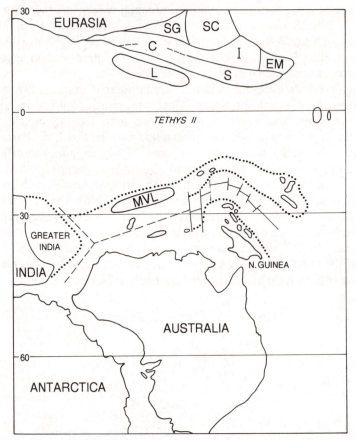

Fig. 7.5 Palaeogeographic reconstruction for Eastern Tethys in the Late Jurassic. SG, Songpan Gangzi accretionary complex; SC, South China; I, Indochina; C, Changtang; EM, East Malaya; L, Lhasa; S, Sibumasu; MVL, Mount Victoria Land. Simplified from Metcalfe (1991).

a remarkable change occurred to shallow marine biotas of Sibumasu in the Mid Permian, with younger faunas having strong affinities with South China and Indochina, being of Tethyan type. This implies that Sibumasu had already rifted from Gondwana. Palaeomagnetic evidence indicates that the final suturing of mainland South-east Asia to Eurasia probably occurred in Early to Mid Jurassic times (Fig. 7.5).

With regard to Tibet, both ammonite and bivalve evidence indicates that the Himalayan zone in the south was part of the southern margin of Tethys in Early and Mid Jurassic times, with strong biogeographic affinities to East Africa and Madagascar. The more limited evidence from northern Tibet, on the other hand, is consistent with a Eurasian position (Westermann

1988; Wen 1986). The Lhasa Block finally collided with the Changtang and Sibumasu terranes in Late Jurassic to Early Cretaceous times, along the Banggong Suture (Dewey 1988).

BIOGEOGRAPHY OF ORGANISMS ON THE PANGAEAN SUPERCONTINENT

Cosmopolitan taxa dominate all Early Mesozoic tetrapod assemblages (Cox 1973, 1974; Rage 1988; Shubin and Sues 1991). The dicynodont reptile *Lystrosaurus* is known from the Lower Triassic of South Africa, Antarctica, India, China, and Russia. Other dicynodonts, the Kannemeyeridae, are found in Lower Triassic deposits only in South Africa, South America, and Australia but increased their range later in the period. The Late Triassic also saw a dispersal across Pangaea of prosauropod dinosaurs, known from North and South America, Europe, China, South and East Africa, India, and Australia. Such cosmopolitanism clearly indicates the lack of either physical or climatic barriers to widespread dispersal in the Pangaean supercontinent.

Shubin and Sues (1991) undertook a rigorous analysis of Early Mesozoic tetrapods, utilizing presence–absence data for families in major, well-documented assemblages. The data were scored in a matrix that was subsequently transformed with a variety of similarity coefficients, and cluster analysis performed on the transformed data. It was acknowledged that a number of factors, such as age, geographic context, palaeoecological factors, and taphonomic biases, may have influenced the degree of similarity of the assemblages.

The Early Triassic was dominated by therapsid reptiles, and very little endemism is recognizable. By the Late Triassic, characterized by advanced cynodont synapsids, dinosaurs, phytosaurs, and rhynchosaurs, Laurasian and Gondwanan groups can be distinguished, but it is thought that these may reflect differences of age rather than geographic separation, because global correlation of the non-marine deposits in which the fossils are found is still imprecise. The Early Jurassic faunas are extremely homogeneous, and no endemism at family level can be recognized, unlike for the Late Triassic.

With regard to the later Jurassic, the two richest dinosaur faunas are from the Morrison Formation of the Western Interior of the United States and the Tendaguru Beds of Tanzania, both Kimmeridgian–Tithonian in age. The faunas have many similarities to each other, and Charig (1973) considers that there is rather tenuous evidence of a homogeneous fauna extending from North America into Europe and Africa. *Brachiosaurus* and *Barosaurus* are two genera in common to North America and East Africa, and Galton (1977) adds a third genus, the hypsilophodontid *Dryosaurus*. A land connection is clearly implied, but it remains uncertain whether it was via western Europe or Central America. The problem is that sea level was higher in the Late Jurassic than earlier in the Mesozoic (Fig. 4.1) and large parts of the purported land routes were flooded by epicontinental sea.

At times of short-term sea-level fall such sea could regress from the continents, and this could explain why several genera of Morrison Formation lizards have also been found in the Late Jurassic of England and Portugal (Prothero and Estes 1980), implying a terrestrial migration route across the North Atlantic region.

A more serious problem concerns the possible land link between Laurasia and Gondwana in the Late Jurassic. At this time, as a consequence of the initial opening phase of the Atlantic and associated transcurrent movement between the Iberian Peninsula of Europe and North Africa, a narrow zone of deep oceanic crust has been inferred between these regions (Dercourt *et al.* 1986). This clearly would not have permitted cross migration of dinosaurs even at times of low sea-level stand.

Galton (1977) favoured a Cental American land connection. This theory receives some support from the discovery in Patagonia of a Callovian–Oxfordian dinosaur fauna showing quite strong affinities to other continents (Bonaparte 1979). However by Oxfordian times an oceanic connection had evidently been established between the western Tethys and the eastern Pacific, isolating South America from North America. Perhaps the resolution of this particular problem lies in the megashear tectonics of Central America, producing a geometry that would have allowed the existence of a land bridge between the two continents long after the initial opening of the Central Atlantic, perhaps even to the end of the Jurassic, when separation to the south of Honduras must have severed the connection (Schweichert comments on Hallam 1981*a*). According to the reconstructions of the Caribbean region by Pindell *et al.* (1988) there was no deep ocean connection between the Atlantic and Pacific in Oxfordian times, and such a connection was only marginally established as late as the Tithonian. Such reconstructions are certainly consistent with dinosaur distributions.

The Early Mesozoic terrestrial flora is characterized by ferns and gymnosperms. For the Early Triassic three floral regions can be distinguished, the Euramerican, Angara, and Gondwana (Fig. 7.6). The Euramerican region lies within about 15° and 55°N and is thought to characterize a subtropical–warm temperate belt, while the Angara region of Siberia probably signifies a cooler but still temperate climate. The third, Gondwana, region, extending from 30° to more than 60°S, contains some taxa found additionally in the northern hemisphere but also endemics, most notably the seed fern *Dicroidium*, which ranges in age up to the Carnian (Barnard 1973).

By the end of the Triassic and into the Jurassic a more cosmopolitan flora

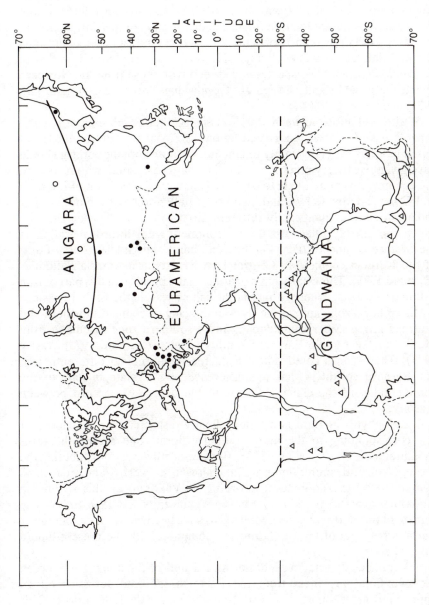

Fig. 7.6 Floristic regions in the Early Triassic. Solid circles, Euramerican flora; hollow circles, Angaran flora; triangles, *Dicrodium* flora. After Barnard (1973).

had become established, typified by diptoridacean and matoniacean ferns, pteridosperms, and various gymnosperms including conifers, Caytoniales, Bennetitales, Nilssoniales, and Ginkgoales. The latest Triassic *Dictyophyllum* flora extended from 50°N to 60°S, and includes ferns whose living relatives cannot tolerate frost. The distribution of this flora therefore indicates both free land connections between Laurasia and Gondwana and a relatively equable climate. However, the southern floras with *Dictyophyllum* have few species in common with those from the Northern Hemisphere, and signify a separate Gondwanan floristic region (Barnard 1973).

A differentiation between the Jurassic northern and southern hemispheres has also been recognized among the conifers (Florin 1963) but this is unlikely to relate simply to climate, because climatic control should have given rise to a symmetrical latitudinal pattern of zones with respect to the equator. It is more likely to be due to increasing separation of Laurasia and Gondwana by Tethys, and inability of pollen to cross the ocean. The Indian floras are significantly different from those in adjacent Eurasia, which presumably relates to the subcontinent's northward migration in the Cretaceous and Tertiary. On the other hand, the Middle Jurassic floras of the Kerman coalfield in Central Iran have typical Eurasian affinities (Barnard 1973). This clearly has a bearing on the position of this part of Iran within the Tethyan zone. Within Laurasia there was, in the Early Jurassic, a northern floral zone among the ferns embracing Greenland, northern and central Europe, Siberia and Japan, and a southern zone extending from Mexico to the Middle East and southern China (Barnard 1973; Wesley 1973). These zones could reflect a degree of latitudinal differentiation, but Barnard considers that the difference might relate in part to the continentality of the climate, that is to the degree of contrast between seasons.

Cycadophytes in Eurasia are latitudinally restricted to a belt more or less corresponding to the more southerly floral zone among the ferns (Vakhrameev 1991; Wesley 1973). Passing northwards into Siberia, the cycadophyte and conifer floras are much lower in diversity and are instead dominated by ginkgophytes. According to Vakhrameev this signifies a latitudinal climatic gradient in Eurasia significantly less than today. The climate of this northern zone is inferred to have been humid and moderately warm, while that of the southern zone compares with the present humid tropical zone.

Vakhrameev detected a slight northward shift of the boundary between the two floral provinces from the Early to the Mid Jurassic and an appreciably greater northward shift from the Mid to the Late Jurassic. This implies a temperature rise through the course of the period, contrary to what Frakes *et al.* (1992) infer (Fig. 4.1). Frakes *et al.* base their conclusions on the purported presence of Middle and Upper Jurassic glacial dropstones in northern Siberia, but some doubt has been cast on their glacial origin (Hallam 1993). What appears certain is that both the Triassic and Jurassic were relatively equable periods, with no polar icecaps, though seasonal temperature differences on the Pangaean supercontinent could have been considerable.

A further advance in our knowledge of Eurasian floras, based on application of the modern plant biome concept, has been made by Ziegler *et al.* (1993) and this may help to resolve some of the conflicting evidence just referred to. Ziegler *et al.* have undertaken ordination analysis, as widely used by plant ecologists, on plant genera for four Triassic and three Jurassic time intervals. They infer that the floras fall into three main climatically related biomes, the dry subtropical, the warm temperate, and the cool temperate. Vakhrameev (1991) assigned these three floras to tropical, subtropical, and temperate climatic zones, a scale that Ziegler *et al.* consider is too warm throughout and gives the impression that the poles were very warm indeed. Even so, their maps show warm temperate floras above 70^0 latitude in the Triassic and occasionally as high as 60^0 in the Jurassic, with no hint of the cold temperate, arctic, or glacial climates that exist at these latitudes today. There is no hint of a tropical rain forest biome.

MARINE ORGANISMS

Most biogeographic work has been done on the richest and most diverse macroinvertebrate group, the molluscs, essentially ammonites, belemnites, and bivalves.

In the Early Mesozoic, as in the Palaeozoic, a tropical–subtropical low latitude zone can be recognized by the presence of reef-building corals and associated fauna and abundant carbonate sediments. In the Jurassic this zone corresponds essentially with the line of the Tethys seaway, from Mexico and the Caribbean to the circum-Mediterranean countries, the Middle East, Himalayas, and Indonesia. Besides corals and stromatoporoids, characteristic faunal elements include rudists and other large, thick-shelled bivalves, nerinacean gastropods (Wieczorek 1988) and lituolacean foraminifers, together with abundant and diverse dasycladacean algae. In the Triassic the Tethyan seaway did not extend into the Caribbean–Gulf Coast region. Its deposits, which include a sub-stantial proportion of carbonates from southern Europe to the Himalayas, contain in the Middle and Upper Triassic a characteristic assemblage of reef-building corals and calcisponges together with megalodontid bivalves, thought to be the ancestors of the rudists (Hallam 1984*b*).

Higher latitude faunas are not so distinctive. Based on his studies of Triassic deposits on the North American craton and in the Alps, Tozer

(1982) has endeavoured to distinguish Low,Mid, and High Palaeolatitude faunas (LPL, MPL, and HPL respectively). LPL faunas are characterized by corals, sponges, megalodontids, and the pteriomorph bivalve species *Monotis salinaria*. MPL and HPL faunas have some distinctive molluscan species, with HPL ammonites being much less diverse than MPL. However, in the earliest Triassic (Griesbachian and Dienerian) the ammonite and other molluscan fauna is cosmopolitan (cf. Kummel 1973) and LPL, MPL, and HPL faunas cannot be distinguished. As regards the later Early Triassic, LPL can be distinguished in the Early Smithian and Early Spathian but not late in these ages. Using conodonts, Matsuda (1985) can distinguish, within an Early Triassic Tethys Realm, a Tethys Province characterized by high faunal diversity, and a Peri-Gondwana Province (Kashmir, Pakistan Salt Range, Nepal, West Australia) with low diversity.

Among other groups, some Late Triassic (Norian) brachiopods have a broadly Tethyan distribution, namely the Halorellinae, most notably the genus *Halorella* (Ager 1973). Kristan-Tollman (1988) has established that there is a remarkable uniformity of benthic organisms even at species level throughout the Tethyan region from the Alps to Indonesia, as can be demonstrated for groups as varied as algae, crinoids, holothurians, and ostracods.

As regards the Jurassic, it has been widely assumed that the well-known Tethyan–Boreal provinciality was controlled by the obvious latitude-related factor, temperature. The matter is likely to be more complicated and warrants some detailed attention.

Tethyan–Boreal provinciality in the Jurassic

Both ammonites and belemnites allow the distinction of Tethyan and Boreal realms for most of the Jurassic, principally the Middle and Upper Jurassic (Hallam 1975). Ammonites are more diverse and morphologically varied and have received much more attention than the belemnites, principally because their high temporal turnover allows refined biostratigraphic subdivisions to be made.

The Boreal Realm occupied the northern part of the Northern Hemisphere and is best defined by the distribution of the following ammonite families or subfamilies.

VOLGIAN	Craspeditidae
	Virgatitidae
	Dorsoplanitidae
KIMMERIDGIAN	Aulacostephaninae
	Cardioceratinae
CALLOVIAN	Kosmoceratidae
	Arctocephalitinae

BATHONIAN	Arctocephalitinae
BAJOCIAN (UPPER)	Arctocephalitinae
PLIENSBACHIAN	Amaltheidae
	Liparoceratidae

In a general way the southern limit of these families corresponds in Europe approximately with a zone through southern Europe, and in the North Pacific region with a zone through northern California and between Japan and Siberia, although the boundary here is complicated by the existence of displaced terranes, as will be discussed later. The boundary is a gradational one and consequently there can be no general agreement on its precise location at a given time. Moreover, the boundary fluctuated in the course of time (Fig. 7.7). Thus in the Late Callovian and Early Oxfordian boreal ammonite genera reached far into southern Europe, while subsequently in the Late Oxfordian the boundary retreated back into northern Europe. (This is more of a boreal retreat than a Tethyan spread, as Arkell (1956) thought; see Cariou 1973). Late Bajocian and Bathonian boreal ammonites are confined mostly to the Arctic regions, although they extend into the North American Western Interior. A distinctive boreal ammonite fauna first became clearly established in Pliensbachian times but was temporarily lost because of extinction of the Amaltheidae. Distinctness was renewed in the Mid Jurassic, reaching a maximum in the Bathonian, subsequently to become blurred somewhat in the Mid and Late Oxfordian, and then set in once more to become extreme at the close of the period, in the Tithonian and Volgian.

There are indications of north–south provinciality. Thus Cariou (1973) distinguished a Boreal Province in the Callovian and Oxfordian in the Arctic, characterized by the dominance of Cardioceratidae, from a Sub-boreal Province, where kosmoceratids dominate in the Callovian and abundant perisphinctids occur in the Oxfordian. A similar provinciality is recognized for the Early Volgian by Zeiss (1968).

Ammonites of the Tethyan Realm, occupying the rest of the world, show greater diversity. The phylloceratids and lytoceratids have long been considered as characteristics of the old Tethys Ocean, which does not mean that they are abundant in all Tethyan faunas, nor that they do not occur sporadically in the Boreal Realm. The Oppeliidae and Haploceratinae are other characteristic groups from Mid Jurassic times onwards. Besides these, the following families and subfamilies have a predominantly or exclusively Tethyan distribution.

TITHONIAN	Berriasellidae
	Spiticeratinae
	Virgotasphinctinae

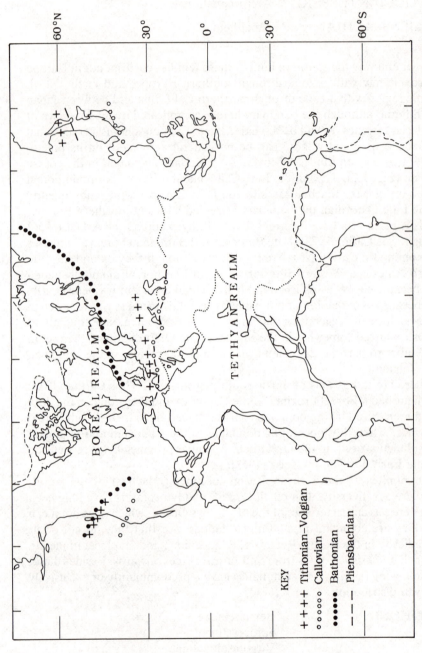

Fig. 7.7 The approximate boundaries of the Tethyan and Boreal Realms at different times in the Jurassic. After Hallam (1975).

KEY
+ + + + Tithonian–Volgian
o o o o o o Callovian
●●●●● Bathonian
– – – Pliensbachian

KIMMERIDGIAN	Aspidoceratinae
	Ataxioceratinae
	Virgatosphinctinae
	Streblitinae
	Simoceratidae
	Taramelliceratinae
OXFORDIAN	Aspidoceratinae
	Ochetoceratinae
	Taramelliceratinae
	Peltoceratinae
CALLOVIAN	Aspidoceratinae
	Reineckiidae
	Hecticoceratidae
	Macrocephalitidae
	Sphaeroceratidae
BATHONIAN	Clydoniceratidae
	Tulitidae
	Zigzagiceratidae
	Morphoceratidae
BAJOCIAN	Hammatocratidae
	Leptosphinctinae
TOARCIAN	Hammatoceratidae
	Bouleiceratinae
PLIENSBACHIAN	Hiloceratidae
	Dactylioceratidae
SINEMURIAN and HETTANGIAN	Ectocentritidae
	Juraphyllitidae

The Tethyan faunas are dominantly cosmopolitan in distribution, even at the species level, but there are more indications of provinciality than in the case of the Boreal Realm, a fact which may be partly related to the much greater geographic extent of the Tethyan Realm. For example, attempts have been made in Europe to distinguish Submediterranean and Mediterranean Provinces but the differences may at least in some cases be related to depth of sea, with for example phylloceratids and lytoceratids occupying deeper water zones than oppeliids and perisphinctids (Ziegler 1967).

The Tethyan–Boreal provinciality among the ammonites may prove a useful tool in helping to unravel the complicated tectonics of the Cimmeride–Alpide orogenic belt of southern Eurasia. Thus Thierry (1988) points out the Caucaso–Dobrodgean Furrow was an oceanic area, according to some palaeogeographic reconstructions, separating Dobrodgea,

Crimea, and the Greater Caucasus in the north from the eastern Pontides and Lesser Caucasus in the south. This deep basin may have acted as an effective barrier preventing Boreal ammonite expansion to the south in the Callovian. On the Russian Platform the faunas that got to the Greater Caucasus never reached the Lesser Caucasus. The Lesser Caucasus and Pontides have Tethyan taxa very like those of Sicily, and no Boreal taxa. However, Sengör and Yilmaz (1981) consider that in the Mid Jurassic this furrow (their 'Palaeotethys') was closed, but if this were the case boreal ammonites would surely have reached the Pontides.

Belemnite provinciality developed late in the Early Jurassic (Toarcian) when distinct Arctic taxa are first recognized (Doyle 1987). In the Mid Jurassic Boreal and Tethyan Realms are apparent, the Boreal Realm being divisible into informal Boreal–Atlantic and Arctic provinces (Doyle 1987). The Boreal Realm was dominated by the Cylindroteuthidae, the Tethyan Realm by Belemnopseidae and Duvaliidae, with *Hibolithes*. The Gondwana shelf seas were dominated by *Belemnopsis*, with *Hibolithes*. Separate Mediterreanean, Ethiopian, and Indo-Pacific Provinces, have been distinguished (Stevens 1973a; Doyle 1987).

In the Early Jurassic the xiphoteuthid belemnoid *Atractites* is quite common in the Tethyan Realm of southern Europe, but absent from the Boreal Realm.

Among the bivalves, which are the most abundant and diverse Jurassic macroinvertebrates, a distinctive Tethyan–Boreal provinciality is more difficult to recognize (Hallam 1977a). Certainly there are some taxa which are characteristically Tethyan in distribution, including rudists such as *Diceras* and a number of other, mainly large, thick-shelled forms, and also the peculiar Early Jurassic genus *Lithiotis*, but not many which are characteristically Boreal. The pteriomorph *Buchia* is perhaps the best known example, extremely abundant in the Upper Jurassic of high northern palaeolatitudes, as is *Plicatula* (*Harpax*) in the Lower Jurassic and *Arctotis* in the Middle Jurassic. A few other possible examples are cited by Damborenea (1993). However, none of these taxa are confined to the Boreal Realm, and they will be discussed again when the possibility of bipolar distributions is considered.

The only other groups where some possible Tethyan–Boreal provinciality has been proposed are brachiopods (Ager 1973) and foraminifers (Gordon 1970). For neither group is the situation as clear-cut as for the ammonites and belemnites.

Among the brachiopods the Pygopidae are the most typical Tethyan forms, probably living in conditions of very quiet water and low food supply (Manceñido 1993). In the Lower Jurassic certain genera, e.g. *Pisirhynchia*, are confined to the peri-Adriatic region but other elements of this fauna, *Prionorhynchia* and *Cirpa*, are more widely distributed, extending into Transcaucasia, southern Spain and even as far as south-west England. In the Pliensbachian of Europe such genera as *Tetrarhynchia, Gibbirhynchia, Lobothyris*, indentate *Zeilleria* and ribbed *Spiriferina* have a predominantly boreal, or at least extra-Mediterranean, distribution, in so far as they are they are absent from Ager's two inner circum- and intra-Mediterranean belts, though elements of the fauna do occur in Spain and Morocco. Only *Grandirhynchia* has a strictly boreal distribution, being apparently confined to Scotland, East Greenland, and north-east Siberia.

The shallow water carbonate belt on the southern side of Tethys, from the circum-Mediterranean region to the Middle East, contains a characteristic foraminiferan fauna of arenaceous forms with complex internal structure, notably the Lituolidae, Pavonitidae, and Dicyclinidea. Typical genera include *Orbitopsella* and *Lituosepta* (Lower Jurassic), *Kilianina, Meyendorffina*, and *Orbitammina* (Middle Jurassic) and *Kurnubia* and *Pseudocyclammina* (Upper Jurassic). Specifically boreal genera are, as with the bivalves and brachiopods, less easy to distinguish, but Europe north of Tethys is characterized by nodosariids and many simple-structured arenaceous genera, with episodic occurrences of genera such as *Epistomina* and *Ophthalmidium* from Mid Jurassic times onwards.

Possible causes of Tethyan–Boreal provinciality.

Since the pioneer work of Neumayr (1883) the most popular interpretation of this provinciality involves temperature, with boreal faunas signifying cooler waters (Hallam 1975). This is based on the fact that the provincial boundary is broadly related to latitude and the higher diversity ammonite faunas of the Tethyan Realm are associated, in other facies within the Alpine belt, with corals and other groups suggesting tropical or subtropical conditions. At least two categories of objection can be raised to this conventional view.

1. If an actualistic comparison is to be made with the present day, there should be a significant reduction in diversity from the tropics to the poles. While there is indeed some diversity reduction among the ammonites from the Tethyan to the Boreal Realm within the best studied region, Europe, this is not the case for the belemnites. The diversity of Doyle's (1987) Boreal–Atlantic Province is actually lower than that of his more northerly Arctic Province.

Although the bivalves do not exhibit clear Tethyan–Boreal provinciality they are a valuable group to be used for a diversity test, both because of their high abundance and diversity in the Jurassic and because of good data for the present day. According to Stehli *et al.* (1967) the diversity of Recent bivalves shows a significant reduction as latitude increases away from the tropics, both at the generic and specific level. Across the 40° range of latitude separating Morocco from Greenland, generic diversity in the present Atlantic is reduced approximately threefold, and even more over a similar range in the Pacific. These changes are so pronounced that, even in a more equable climate than today, a northward reduction in diversity seems essential to support any temperature control hypothesis. Hallam (1972) conducted a study of Pliensbachian and Toarcian bivalves of Morocco, Iberia, England, and Greenland. On a regional scale, and making due allowance for facies, bivalve diversity tends to increase from the Tethyan to the Boreal Realm. The bivalve data, like the belemnite data, are thus incompatible with a simple temperature control hypothesis. Another relevant study concerns a group of planktonic microorganisms, the dinoflagellates. Smelror (1993) undertook a biogeographic study by comparing the Bathonian to Oxfordian microflora of the Arctic, north-west Europe, and circum-Mediterranean regions. Few species are endemic to the Tethyan and Boreal Realms but different taxa dominate in the various regions. The highest diversity is in north-west-Europe and falls both to the south and north.

2. If the Tethyan–Boreal provinciality relates to temperature there should be latitudinally oriented bilateral symmetry, with an Austral Realm in the southern hemisphere corresponding to the Boreal Realm in the northern hemisphere. No such realm is recognizable among either the ammonites or the belemnites, the only two groups which clearly indicate a boreal provinciality. There are Jurassic marine invertebrates showing bipolar distribution: Crame (1986) and Damborenea (1993) make a claim for a few genera of bivalves, such as *Buchia*, *Retroceramus*, *Arctotis*, *Palmoxytoma*, and *Plicatula* (*Harpax*). To take the much discussed case of *Buchia*, most occurrences are north of 30°N palaeolatitude, and in the Himalayas, located south of 30°S in Late Jurassic time. *Buchia* is common in West Antarctica, becoming less common in Andean South America.

However, *Buchia* and the other bivalves have a circum-Gondwana distribution, which may be more significant than latitude, and *Buchia* occurs commonly in North American Cordilleran terranes as far south as Northern Mexico. Belemnites of Doyle's (1987) Arctic Province also occur in California, associated with Tethyan belemnites. Furthermore, the facies associates of the assemblages of abundant *Buchia* in the Tibetan Himalayas suggest relatively deep-water conditions, because they occur with phylloceratid and lytoceratid ammonites (Li and Grant-Mackie 1993). Deeper water is normally cooler water, and it is possible that some form of temperature control was operative for *Buchia*, but this evidently reflects more than a simple latitudinal relationship.

Damborenea (1993) cites *Palmoxytoma* as an Early Jurassic bivalve with a bipolar distribution but it also occurs in the Northern Calcareous Alps of Austria, in the heart of the Tethyan Realm.

Despite these criticisms, many Jurassic marine organisms, like those at other times, were confined to low latitudes, but the critical controlling factor at a time when the tropics to poles temperature gradient is likely to have been appreciably less than today, may perhaps have involved seasonal changes in illumination and/or solar radiation (Reid 1973).

The likeliest cause of Jurassic Tethyan–Boreal provinciality involves palaeogeography and associated environmental stability (Hallam 1975, 1984b; Fürsich and Sykes 1977; Doyle 1987). The provinciality among the ammonites is most clearly defined at times of either regional or global regression, in the Bathonian and Tithonian/Volgian. Restriction of a seaway connecting the Arctic to Tethys via western Europe, due to uplift in the North Sea region, was almost certainly a major factor in Bajocian–Bathonian times. In the later Callovian and Oxfordian, there was an increase in what has been called 'dispersicity' of dinoflagellates, which coincides with the disappearance of land barriers and the establishment of new open marine seaways, so that regional differences in the composition of assemblages became less prominent (Smelror 1993). Although the 'Arctic' belemnites *Arctoteuthis* and *Lagonibelus* could, in the Pacific, reach as far south as California in Tithonian times, they were inhibited from entering western Europe by geographic resrictions in a region of shallow seas and intervening land masses (Doyle 1987; cf. Fürsich and Sykes 1977). Perhaps most tellingly of all, the Tethyan–Boreal provinciality among ammonites and bivalves broke down in the Late Cretaceous, coincident with a time of drastic palaeogeographic change associated with the opening of the North Atlantic.

Pacific faunas.

Although many bivalve taxa are cosmopolitan there are some that are not recorded from Europe, Africa, or western Asia. Fig. 7.8 illustrates the distribution of some Upper Triassic bivalve genera and species that characterize the Pacific margins. Some, such as *Minetrigonia* and *Monotis ochotica* occur on both sides of the Pacific, while others, such as *Septocardia* and *Monotis subcircularis* (east) and *Tosapecten* and *Monotis zabaikalia* (west) occur only on one side. For a long time New Zealand workers distinguished a Maorian Province for the south-western Pacific in Late Triassic and Early Jurassic times, based on the apparent restriction to New Zealand and New Caledonia of such genera as the brachiopods *Clavigera* and *Rastelligera* and the bivalves *Praegonia*, *Maoritrigonia*, and *Kalentera* (Stevens 1980). A number of these taxa, such as *Clavigera* and *Kalentera*, have since been found in Chile, throwing into question the existence of a separate Maorian Province (Damborenea and Manceñido 1992).

The Lower Jurassic of the east Pacific borders of North and South America is characterized by the distinctive pectinid genus *Weyla*, which occurs in great abundance. Although known also from the Pliensbachian of Europe and the Toarcian of Madagascar and the Himalayas, one could almost define the east Pacific region as a *Weyla* province, it is so

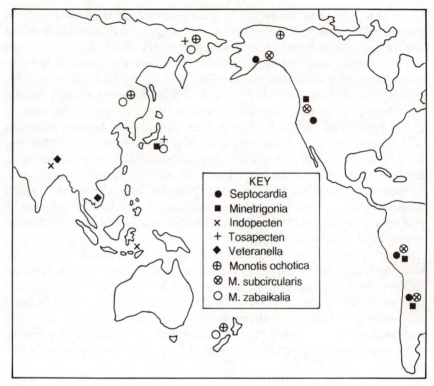

Fig. 7.8 Important 'Pacific' elements in Late Triassic bivalve faunas. After Hallam (1981c).

regionally distinctive; a point which possibly reveals a limitation of using similarity coefficients based on presence–absence data. Apart from *Weyla*, *Lupherella* is the only other bivalve genus apparently confined to the east Pacific (Damborenea 1993).

Northern Chile and Argentina contain other Lower Jurassic bivalve genera, such as *Lithiotis* (also recorded from Oregon) and *Gervilleoperna* suggesting close links with Tethys. According to Damborenea and Manceñido (1992) Middle and Upper Jurassic strata in Argentina contain inoceramids strikingly similar to inoceramids of the same age in New Zealand. Where the faunas of the two regions differ, these authors believe that it is likely to be the consequence of facies differences. Thus, in the Early Jurassic most of the characteristic taxa of the New Zealand faunas have their counterparts in the outer shelf facies in the southern Andes. The other, inner shelf, bivalve elements such as *Weyla*, *Lopha*, *Ctenostreon*, and trigoniids are not present in New Zealand because the appropriate facies is not preserved. The only conspicuous New Zealand element not found in the

Andes is *Pseudaucella*. The inference that suitable neritic habitats on the Gondwana margins permitted free faunal exchange is clearly correct, but whether the information is sufficient to allow the definition of a distinctive 'Palaeoaustral Realm' is open to question.

With regard to east–west Pacific links in the Early Jurassic, the bivalve genera *Otapiria* and *Pectinula* are found in the southern Andes and the north-west and south-west Pacific margins (*Pectinula* only in Japan) and nowhere else (Hallam 1977*a*), although *Otapiria* is also known from a wider region in the Upper Triassic, extending as far as the western Tethys (Austria).

Ammonite taxa tend to have more cosmopolitan distribution than bivalves but some Pacific endemism has been reported for Jurassic forms. Moreover, Pacific faunas may be dominated by different taxa than contemporary ones in Europe, even when no strict endemism can be discerned (Westermann 1981; Hillebrandt *et al*. 1993).

In the Hettangian, *Badouxia* appears to be restricted to the Pacific margins of North and South America and north-east Asia. Although *Sunrisites* occurs in the Austrian Alps it is only really common in the Americas. Unlike in Europe, Alsatitinae, not *Schlotheimia*, dominated in the Americas in the Late Hettangian. In general regional differentiation of ammonite faunas is weak in the Hettangian and Sinemurian.

With regard to the Pliensbachian there are some important differences from Europe. *Oistoceras, Beaniceras, Androgynoceras*, and *Pleuroceras* are not known from the circum-Pacific region and *Fanninoceras* is the most characteristic genus in North and South America, defining an East Pacific Subrealm. For the Toarcian, the familiar European genus *Hildoceras* is not yet reported from the circum-Pacific region, and *Grammoceras* and *Pseudogrammoceras* are unknown from South America where, however, *Phymatoceras* is commoner than in Europe (Hillebrandt *et al*. 1993).

In the Middle Jurassic, some Early Bajocian ammonite genera and subgenera indicate a direct link between the western Tethys and the western Americas but early in the Late Bajocian the entire eastern part of the Pacific margin became an independent province, called the East Pacific Realm by Westermann (1981). The abundant supposed Early Callovian '*Macrocephalites*' and *Indocephalites* of the Andes and Mexico are now recognized as a new Bathonian subfamily, the Eurycephalitinae (Thierry 1976). The Macrocephalitinae appear to be entirely absent from the east and north-west Pacific but are abundant in the south-west Pacific.

Late in the Early Callovian the east Pacific endemism disappeared, but differences with Madagascar and Australasia (mainly New Guinea) remained, because the faunas of those regions include important Tethyan (Macrocephalitidae) and East Tethyan (Eucycloceratidae) groups unknown in the Andes. In the Oxfordian, the similarity between South American and Tethyan ammonites is, for the first time since the Hettangian, higher than between South and North America, but differences persist between South America and Madagascar–New Guinea, because the ammonite fauna of those regions includes the Mayaitidae, a family unknown from the Andes, despite earlier claims (Riccardi 1991).

The first Jurassic belemnoid in the Americas is the xiphoteuthid *Atractites*, which is common in the Pliensbachian and Toarcian in Chile. Aulacocerids are known from the Americas prior to the Toarcian, descended from a rich fauna already established in the cordilleran region in the Triassic. They were replaced in the Toarcian and early Middle Jurassic by true belemnites (Challinor *et al.* 1993). It will be recalled that *Atractites* is quite common in the Lower Jurassic of the Old World Tethys, but absent from the Boreal Realm.

CIRCUM-PACIFIC DISPLACED TERRANES

It is generally accepted that the Pacific Ocean is bordered by displaced terranes, which accreted to the Asian and American continents in post-Mid Jurassic, mainly Cretaceous time (Howell 1985). Since much of the key evidence derives from marine invertebrate faunas of Late Palaeozoic to Jurassic age it is appropriate to review the subject in this chapter. Although it has been claimed that the whole South American Pacific borderland consists of displaced terranes (Tozer 1982) there are good reasons to reject this, and the only accreted segments, in the far north and south, lack invertebrate fossils (Hallam 1986). Accordingly attention will be confined to four regions that have received detailed study, Cordilleran North America, Japan, eastern Siberia, and New Zealand.

Cordilleran North America (Fig. 7.9)
The Cache Creek Terrane of Canada is a narrow, linear belt extending 1500 km from southern British Columbia to the Yukon. It contains Permian foraminifers (verbeekinid fusulines) and Carboniferous corals (pseudopavonids and waagenophyllids) of Tethyan and Asiatic affinity (Monger and Ross 1971; Stevens 1983). This has led to the suggestion that the original location of the terrane components was in the western Pacific, probably in the form of scattered islands or oceanic plateaus (Stevens 1983; Ross and Ross 1985).

In southern Alaska there is a thick succession of Upper Triassic carbonates containing scleractinian corals, spongiomorphs, megalodontid bivalves, and brachiopods, indicative of Tozer's (1982) Low Palaeolatitude fauna. According to Tozer, such faunas occur in four Canadian terranes, from west to east Wrangellia, Alexander, Stikinia, and Quesnellia (Fig. 7.9). Carnian and Norian faunas with scleractinians north of 60°N, as in Wrangellia, suggest some 3000 km northward displacement. Distinctive bivalves common to these terranes, and the Luning Terrane of western Nevada, suggest an original occurrence in the same province. There is agreement with palaeomagnetic data indicating northward displacement for Wrangellia and Quesnellia. The bivalve *Monotis salinaria* in these terranes is another indicator of low latitude Tethyan affinities (Silberling 1985).

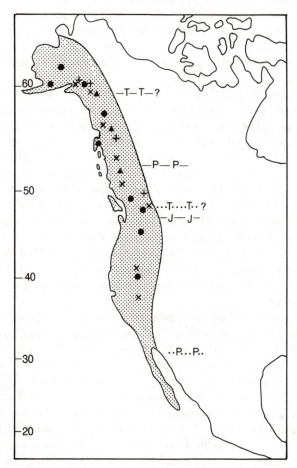

Fig. 7.9 The zone of displaced terranes (stippled) of the North American cordillera, with location of latitudinally anomalous faunas. Tethyan faunas signified by black triangles (Permian), black circles (Triassic), and oblique crosses (Jurassic, Pliensbachian). Boreal faunas of Jurassic (Pliensbachian) signified by vertical crosses. Within the craton, the boundaries of high and middle palaeolatitude faunas in the Permian and Triassic are indicated by continuous line and between middle and low palaeolatitude faunas by dotted line. The Tethyan–Boreal provincial boundary for the Jurassic is also shown. After Hallam (1986).

Tozer (1982) envisages an original location of the cordilleran terranes as a series of volcanic archipelagos in the East Pacific, within 30° of the equator, with some carbonate banks and coralline shoals, and deeper basins in between. As a result of detailed analysis of Triassic bivalves of the Wallowa Terrane in Oregon, Newton (1987) recognizes a high level of species endemism and makes a comparison with the Hawaiian Islands of the present day. There are evidently many similarities, but with the Wallowa Terrane having much stronger North American affinities, implying greater proximity to that continent. This conclusion is in agreement with the palaeogeographic inferences of Tozer but not with those like Stevens (1983) and Ross and Ross (1985) who favour a West Pacific location for at least one terrane.

An East Pacific location of the terranes is also indicated by Lower Jurassic fossils, such as the bivalve *Weyla* and the Pliensbachian ammonite *Fanninoceras*. There are also latitudinal as well as longitudinal constraints, provided by the boreal Lower Jurassic ammonites *Amaltheus* and *Arctoasteroceras*, restricted to the northern part of the cordilleran terranes. The more distal to the craton, the more northerly the displacement. Restoration implies a latitudinal displacement of 500 km for Quesnellia, 1800 km for Stikinia, and 2400 km for Wrangellia (Taylor *et al.* 1984; Smith and Tipper 1986). The most plausible plate tectonic scenario is that as the North Atlantic opened in the Late Mesozoic North America was carried westwards and collided with an archipelago of East Pacific islands and submarine plateaus. Oblique impingement ensured that there was a progressive northward displacement of accreted terranes along a series of strike-slip faults.

An ingenious approach to attempting to infer the approximate longitudinal position of Pacific-derived displaced terranes in North America has been adopted by Belasky and Runnegar (1993). They used the generic diversity of living reef corals together with palaeomagnetic results from Quaternary lavas to calculate the present locations of Easter Island, Hawaii and Tahiti. Their results demonstrate that the eastward decline in coral diversity in the Pacific, established by trend-surface analysis, may be used as a measure of longitude. Western North American terranes, for example Eastern Klamath and Cache Creek, contain only a few Tethyan coral taxa of Permo-Triassic age. This is consistent with an original eastern Pacific location of the terranes, but more studies are required of the diversity of correlative coral-bearing strata in the western circum-Pacific region.

Japan. As with cordilleran North America, Japan is now recognized to be a collage of terranes. Three Permian fusuline 'provinces' or territories are recognized by Ishii *et al.* (1985). The South Kitakami Terrane of north–eastern Honshu has close affinities to north–eastern China and Sikhote Alin, notably the presence of *Monodiexolina*. On the other hand the Akyoski–Omi Terrane of south-western Japan has affinities to South China and South-East Asia, especially in the abundant *Colania*. The Tamba–Mino–Ashio and Chichibu Terranes have *Yabeina* and *Neoschwagerina simplex*, of Tethyan affinities. The Akiyoski–Omi Terrane is thought by these authors to have accreted in the Triassic, the others in the Jurassic to earliest Cretaceous.

The Triassic bivalve fauna of the so-called Inner Zone of Japan, closest to the Asian mainland, is very similar to that of eastern Siberia, with distinctive endemic taxa such as *Tosapecten* not found anywhere else, and virtually identical to that in Sikhote Alin (Kobayashi and Tamura 1984). The same is true for Jurassic bivalves (Hayami 1961). The occasional presence of the boreal ammonites *Amaltheus* (Pliensbachian) and *Kepplerites* (Callovian; Sato, 1962) rules out significant post-Mid Jurassic northward migration of land. On the other hand, in the Outer Zone of south western Japan, the Upper Jurassic Torinosu fauna has a diverse assemblage of reef corals, stromatoporoids, nerineids and Tethyan-type bivalves with strong affinities to South-East Asia, especially the identical species of trigoniids and *Somapecten*, found only here and in Sarawak (Hayami 1984). Similarly, the Upper Triassic *Dicerocardium* is a Tethyan bivalve found in Kyushu (Kobayashi and Tamura 1984). These various fossils imply considerable northward displacement of the relevant terranes in post-Jurassic time. More generally, both the Triassic and Jurassic bivalves of the west Pacific region show stronger affinities to mainland Asia than to the east Pacific, thus excluding the likelihood of significant longitudinal displacements (Hayami 1961, 1984; Kobayashi and Tamura 1984; Westermann 1973).

A revolution in Japanese geology has been brought about by radiolarian stratigraphy. Thus the geological age of the Chichibu Group in southern Japan is not Permo–Carboniferous, as originally thought on the basis of the age of large masses of limestone, but Jurassic (Mizutani and Kojima) 1992). The limestone masses turn out to be exotic blocks, but they cannot be recognized in the field as allochthonous. They are now thought to represent an accretionary complex along the ancient continental margin, with disrupted seamounts and surrounding ocean floor sediments intercalated with much younger clastic material (Figs 7.10, 7.11). The Upper Jurassic Torinosu fauna of Shikoku and Kyushu belongs to part of this subduction complex.

Eastern Siberia.

This region, from Sikhote Alin to the Chukotka Peninsula facing Alaska, is part of a collage of accreted terranes (Fujita and Newberry 1982). Greatest interest attaches to the Koryak Uplands north of Kamchatka, more than 60°N (Fig. 8.2). This has a number of Tethyan faunas, including Permian fusulines (Yancey 1979), the Triassic brachiopod *Spondylospira* together with megalodontid bivalves and corals (Dagys 1993) and Upper Jurassic

OFFSCRAPE–ACCRETION

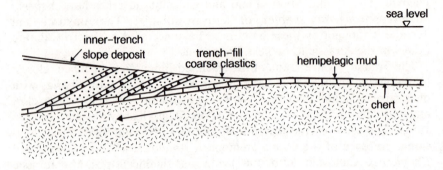

COLLISION–ACCRETION

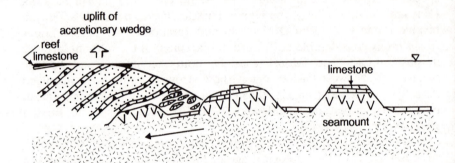

Fig. 7.10 Schematic reconstruction showing two major tectonic stages of the Southern Chichibu Terrane, Japan. After Mizutani and Kojima (1992).

corals (Beauvais 1993). This implies a northward translation of a minimum of 30–35°. The Albian–Cenomanian radiolarian fauna is also thought to be a low-latitude one, probably marking the northern boundary of the temperate–subtropical zone (Vishnevskaya 1992), implying at least some post-Mid Cretaceous movement until accretion.

Recent palaeomagnetic data show that the Upper Mesozoic formations of Kamchatka may also have experienced large-scale latitudinal migration, an inference supported by the Mid Cretaceous radiolaria, whose high diversity is thought to signify tropical palaeolatitudes (Vishnevskaya 1992). The occurrence of rich verbeekinid fusuline faunas 4–8° north of Vladivostok indicates at least 20–25° northward movement of part of Sikhote Alin (Yancey 1979).

New Zealand.

The New Zealand microcontinent is a fragment of Gondwana that separated in Late Cretaceous to Tertiary time, during the opening of the Tasman Sea. A western Palaeozoic foreland consolidated by the Devonian (mainly South Island) can be distinguished from an eastern, younger belt mainly underlain by a basement of greywacke-type sequences in terranes accreted and consolidated during the Mesozoic (Spörli 1987 and Fig. 7.12).

There are rich Permian to Jurassic invertebrate faunas in the Murihiku (or Hokonui) Terrane in the west. Permian brachiopods show strong affinities and comparable diversity to those in eastern Australia (Waterhouse and Bonham Carter 1975) suggesting no great longitudinal or latitudinal separation. A similar conclusion as regards latitude can be drawn for Triassic and Lower Jurassic times, based on distinctive brachiopods and bivalves of the so-called 'Maorian Province' that Stevens (1980) considers a possible

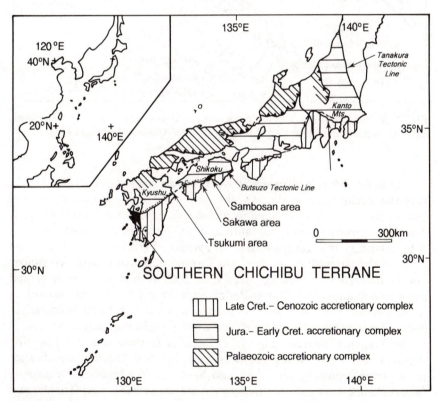

Fig. 7.11 The Southern Chichibu Terrane fringes the southern margin of the Jurassic–Early Cretaceous accretionary complex of South-west Japan. After Mizutani and Kojima (1992).

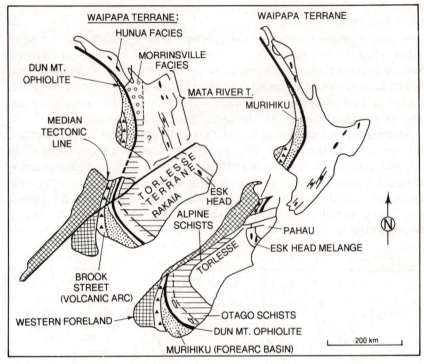

Fig. 7.12 New Zealand basement terranes, left on Mesozoic reconstruction, right on present configuration. Simplified from Aita and Spörli (1992).

austral fauna. More controversially, it has been suggested by Tozer (1982) that the occurrence of the Anisian ammonite *Amphipopanoceras* and Norian nautiloid *Proclydonautilus mandevilli* and Norian bivalve *Monotis* ? *ochotica* signify a moderate latitude in the northern hemisphere.

In contrast, the Tethyan *Neoschwagerina*, *Verbeekina*, and *Yabeina* occur in the Permian of the Torlesse Terrane of North Island. As Yancey (1979) pointed out, this occurrence at 35°S is highly anomalous and suggests a minimum 5–10° post-Permian southward movement. According to McKinnon (1983) the Torlesse Terrane was rafted in to the volcanic arc system by transform faulting parallel to the Gondwana margin.

The Waipapa Terrane (Fig. 7.12) includes fusuline and coral-bearing Permian limestones whose palaeogeographic and tectonic significance have proved controversial. Aita and Spörli (1992) recognize a striking faunal change between two radiolarian assemblages. Upper Triassic and Lower Jurassic chert faunas have strong Tethyan affinities, but Middle and Upper Jurassic argillite faunas have distinctly non-Tethyan features in species composition and diversity. The predominance of taxa of large size, with thick walls, corresponds ecologically to Recent high-latitude cold water faunas where there is marked seasonal variation in productivity. The differences are not thought to be due to facies and latitude is considered much the more likely factor. This conclusion would imply that the cherts were deposited thousands of kilometres away from the argillites before being brought together by sea-floor spreading and accretion. A polar position of the New Zealand foreland in the Late Jurassic (Spörli *et al.* 1989) is thought to be compatible with the palaeolatitude interpretation. Unfortunately a palaeomagnetic test has proved unsuccessful because of strong overprinting.

There is also a considerable contrast between the Upper Jurassic radiolarian faunas of the Waipapa and Murihiku Terranes. The latter has both Tethyan and cosmopolitan species, corresponding to the interpretation from macrofossils (Stevens 1980). This may indicate a complex reshuffling of terranes to their present position, because if the Murihiku faunas are not high latitude, sediments of the terrane must have been deposited some distance from the New Zealand Gondwana margin, which occupied a very high latitude at the time (Spörli *et al.*, 1989).

Turning to longitude, the Norian *Monotis* faunas of the Murihiku and Torlesse Terranes are quite distinct from those of the accreted terranes of the north-eastern Pacific margin (Silberling 1985). This is a strong argument against the breakup of Pacifica (Nur and Ben Avraham 1977).

MARINE CORRIDORS ACROSS WESTERN PANGAEA

In Triassic times the western limits of Tethys were in the present Mediterranean region, so that there was no direct communication with the 'Palaeopacific' or Panthalassa via the Central American region. This poses a problem, because there are many fossils of Tethyan affinities in the American Cordillera. This problem has been addressed by Newton (1988) who rejects the alternative favoured by others such as Stevens (1983) and Ross and Ross (1985) who argue, on the basis of Permian faunal distributions, that cordilleran terranes have migrated from the west Pacific. Newton favours pantropical distributions and her conclusion is supported by the Jurassic faunal distributions discussed in the previous section. Many Tethyan taxa occur over a wide area between the Alps and Japan, and often the species are indistinguishable (Kristan-Tollmann 1988) implying free migration at least of larvae. Crossing of the Pacific would be facilitated by the utilization of the 'island stepping stones' proposed by Tozer (1982), which subsequently became accreted to the advancing North American continent. Newton also points out that, by taking account of data on modern Pacific molluscs that indicate unequivocal west to east migrations, an equatorial counter current or an intense, jet-like undercurrent could have facilitated larval transport. In favour of some current dispersal, as also favoured by Tollmann and Kristan-Tollmann (1985), is the fact that crustacean coprolites similiar to those described by Kristan-Tollmann (1988) have been recorded by Senowbari-Daryan and Stanley (1986) from Peru, which is unlikely to have been composed of far-travelled terranes (Hallam 1986).

In the Jurassic several trans-Pangaea corridors became established as the supercontinent began to break up and sea level rose. These are indicated in Fig. 7.13, as A_1, A_2, and A_3. A1 was an epicontinental seaway between the Arctic and Tethys between Greenland and Norway, termed the Viking Corridor by Westermann (1993). A_2 is a Central Atlantic seaway (the Hispanic Corridor of Smith 1983), initially epicontinental and later fully oceanic, and A3 a seaway linking East Africa with the southern Andes. As with A_2, it was initially an epicontinental seaway that invaded a zone of stretched and thinned continental crust that was subsequently converted into an oceanic strait as East and West Gondwana began to separate. In addition to these trans-Pangaea routes, organisms could of course have migrated round the periphery of the supercontinent by routes P_1 (boreal) and P_2 (austral).

Viking Corridor.

According to Doré's (1991) tectonic analysis, a continuous seaway through the Atlantic rift system was not established until the end of the Early Jurassic

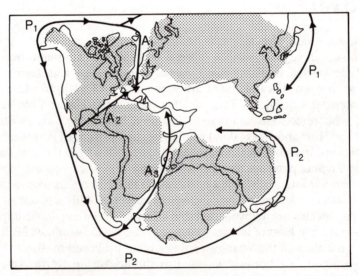

Fig. 7.13 Jurassic marine corridors across Pangaea (A_1–A_3) and routes around Pangaea margins (P_1, P_2). Stippled areas signify land. After Hallam (1983c).

(Toarcian). However, the distribution of amaltheids indicates that it was already in existence in the Late Pliensbachian, because palaeogeographic data preclude any migration of these boreal ammonites either west of Greenland or east of Scandinavia, and migration via Tethys was impossible (Hallam 1977*b*).

Early in the Mid Jurassic uplift centred on the central North Sea restricted free communication at least for ammonites (Doré 1991) and resulted in a marked separation of the Boreal and Tethyan Realms in the present North Atlantic region (Callomon 1984). It reopened only at about the Bathonian–Callovian boundary, allowing free migration of boreal ammonites into western Europe (Westermann 1993). The freedom of faunal migration increased in Late Callovian–Oxfordian times, as indicated by the distribution of dinoflagellates (Smelror 1993) but some restrictions upon entry into the western European region persisted (Doyle 1987). Intense tectonic activity, involving rifting and the creation of half-graben structures, in the latest Jurassic, together with a sea-level fall in the Mid Volgian, led to the re-establishment of marked Tethyan–Boreal ammonite provinciality (Doré, 1991).

Hispanic Corridor.

Newton (1988) has pointed out that the lack of marine Lower and Middle Jurassic rocks in most of Central America argues against a direct marine connection between the Pacific and Tethys, but the available evidence does not preclude the existence of a narrow epicontinental seaway and there is in fact good biogeographic evidence of such a connection.

The earliest clear indications are in the Pliensbachian. The bivalve genus *Weyla*, so common in the North and South American cordilleras, also occurs more rarely in the Iberian Peninsula at that time but nowhere else (Damborenea and Manceñido 1979). *Lithiotis* is known only in the Mediterranean countries, Peru, Chile, and Oregon. Among the ammonites, *Dubariceras* is similarly restricted geographically and *Fanninoceras* is probably derived from the closely related *Radstockiceras*, which is found only in the Sinemurian of Europe (Smith and Tipper 1986). The calcisponge *Stylothalamia* is a western Tethyan genus also found in Peru (Hillebrandt 1981).

While some restriction to free communication persisted until the Early Callovian there was an episode of relatively free intermigration in Toarcian to Early Bajocian times, associated with a major rise in sea level. One of the most significant effects was the spread into north-west Europe of bivalve and other taxa previously restricted to the Pacific margins, to occupy ecological niches vacated by a mass extinction event (Hallam 1983*c*).

As noted in the section on Pacific faunas, an independent East Pacific province developed among ammonites in the Late Bajocian to Early

Callovian time interval, as a consequence of regional uplift restricting communication (Westermann 1981, 1993). Mid Jurassic costate terebratulid brachiopods confined to Morocco and Mexico are good examples of disjunct endemism indicating a direct connection at some time during the subperiod (Ager and Walley 1977) and Westermann's results indicate intermigration in Aalenian to Early Bajocian time. By the Oxfordian free communication had become established on a surer footing, as Africa pulled away from eastern North America and a true oceanic strait opened up (Fig. 7.14). This is also clear from the biogeographic evidence of ammonites, with such genera in Mexico and Peru as *Hybonoticeras*, *Pseudolissoceras*, and *Proniceras* indicating a direct Mediterranean link (Enay 1973). An explanation for the close affinities of Late Jurassic East African and North American dinosaurs possibly due to a transoceanic land connection was discussed earlier.

There is also relevant evidence from the distribution of marine reptiles. According to Gasparini (1993) these were able to disperse along the Hispanic Corridor in the Early and Mid Jurassic. East Pacific and Mediterranean genera and even species show close affinities throughout the period. West Pacific forms are related to those in the Mediterranean region only at family level.

The corridor linking East Africa with the southern Andes.

On the basis of the purported occurrence of the ammonite family Mayaitidae in Argentina, Hallam (1977b) postulated that an epicontinental seaway including the Mozambique Channel had become established as early as the Oxfordian. However, Riccardi (1991) has pointed out that the mayaitid claim is based on a misidentification and that no marine sediments of this age occur either onshore or offshore of South Africa. By Tithonian times it is evident that a seaway had become established. This is shown by organisms of restricted geographic distribution (mainly East Africa–Madagascar and the southern Andes), providing further cases of disjunct endemism. These include ammonites (Enay 1973), belemnites (Mutterlose 1986), and such bivalves as certain trigoniids (Hallam *et al.* 1986) and *Megacucullaea* (Riccardi 1991).

THE INFLUENCE OF SEA LEVEL

It has already been noted that the greater intermigration of bivalves between the Americas and Europe in the Toarcian to Early Bajocian time interval coincides with a rise of sea level. That this phenomenon is a more general one is clearly illustrated in Fig. 7.15, which demonstrates an inverse correlation between the incidence of endemism and degree of inundation of the continents during the Jurassic. Different orders of

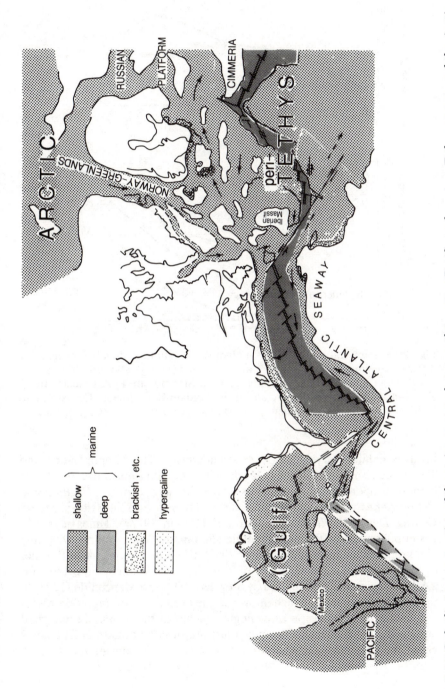

Fig. 7.14 Oxfordian reconstruction of the major seaways between the eastern Pacific and western Tethys oceans and the Arctic Sea. After Westermann (1993).

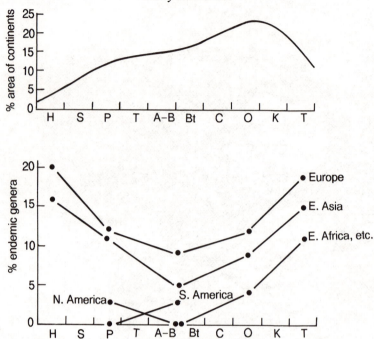

Fig. 7.15 Temporal changes in endemism of Jurassic bivalve genera for different regions (lower graphs) and percentage of continents covered by sea through Jurassic time. H, Hettangian; S, Sinemurian, P, Pliensbachian; T, Toarcian; A-B, Aalenian–Bajocian; Bt, Bathonian; C, Callovian; O, Oxfordian; K, Kimmeridgian; T, Tithonian. After Hallam (1977a).

bivalves exhibit varying degrees of endemism. The Hippuritoidea and Trigonioida show the highest levels while the Pterioda have much the highest proportion of pandemics. They include *Bositra*, *Buchia*, and *Retroceramus*, which compare in cosmopolitanism with the Triassic genera *Claraia*, *Daonella*, *Halobia*, and *Monotis* (Hallam 1977a). Times of greater cosmopolitanism of the nektobenthic ammonites (Westermann 1993) and planktonic dinoflagellates (Smelror 1993) also tend to coincide with times of high sea-level stand. Influxes of Tethyan Jurassic faunas into the New Zealand region are also related by Stevens (1990) and Manceñido (1993) to times of high sea level, which improved the availability of migration routes and caused the expansion of ecological niches in the nearshore and shelf environments. All these examples illustrate the point made in Chapter 2 about the influence of sea level on marine cosmopolitanism.

8

Late Mesozoic

The Cretaceous Period was a time of considerable geological activity associated primarily with the disintegration of Pangaea and a considerable increase in volcanism, which had significant biogeographic consequences. As in previous chapters attention will be directed initially to the framework of major geological events before considering organic distributions.

MAJOR GEOLOGICAL EVENTS

The most up-to-date plate tectonic reconstructions for the Cretaceous are those of Scotese *et al.* (1988), which have utilized an interactive computer graphics method. The three dimensional capabilities of this method allow the rotation and manipulation of plate outlines in 'real time'. Two major phases of plate reorganization are recognized, in the Mid Cretaceous (95 Ma) and latest Cretaceous (65 Ma). The synchroneity of these phases across the world indicates that plate motions are interconnected and suggests to the authors that the reorganizations are triggered by the subduction of major ridge systems, or by the elimination of subduction zones due to continental collision. Furthermore the implication is that 'slab pull' is the dominant plate-tectonic mechanism, with oceanic spreading centres passively following lines of stress emanating from ocean trenches. Fig. 8.1 gives Scotese *et al.*'s global reconstruction for the Late Cretaceous, indicating the areas of new ocean floor.

During the Mid Cretaceous, starting in the earliest Aptian, volcanic eruptions on a massive scale took place, registering an extraordinary upwelling of heat and deep-mantle material. Basalts were initially erupted beneath the Pacific basin and created most of the oceanic plateaus of the present-day western Pacific (Winterer 1991). Eruptions from these mantle upwellings spread to other oceans and sea-floor spreading rates increased. The overall effect was to increase the Earth's ocean crust production by 50–100 per cent during the time interval 125 to 80 Ma. Since this time substantially coincides with a long episode of constant normal geomagnetic polarity (the so-called Quiet Zone of the ocean floor) Larsen (1991) has proposed a superplume model whereby the removal of large quantities of heat and deep-mantle material stopped the reversal process of the Earth's magnetic field.

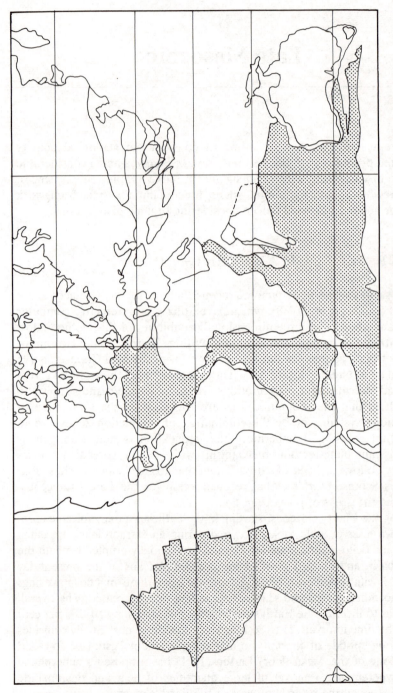

Fig. 8.1 Plate reconstruction for the Late Cretaceous (84 Ma). Stippled areas signify Jurassic and Cretaceous ocean floor produced by sea-floor spreading in Atlantic, Pacific, and Indian Oceans. Simplified from Scotese et al. (1988).

There are a number of important geological consequences of this superplume activity. Extensive outgassing of CO_2 would have led both to climatic warming, due to the greenhouse effect, and also to substantial oceanic dysoxia and anoxia, leading to extensive black shale and petroleum formation (Larsen 1991; Winterer 1991). The rise of oceanic plateaus, together with the increased volume of oceanic ridges due to both an increase in length as the Atlantic and Indian Ocean ridge systems extended, and to the increased sea-floor spreading rate, leading to more extensive hot and therefore buoyant ridges, caused the greatest eustatic rise of sea level since the Early Palaeozoic (Fig. 4.1). The best estimates for the amount of sea-level rise approximate to 250 m. One important consequence for the continents is that up to 40 per cent of their area was flooded (Hallam 1992).

Opening of the Atlantic and Indian Oceans

Sea-floor spreading in the South Atlantic started in the Valanginian, with the ocean between South Africa and Argentina beginning to open about 130 Ma (Scotese *et al*. 1988). It subsequently spread northwards and all connections between Africa and South America were severed by early in the Late Cretaceous, between 95 and 80 Ma (Sclater *et al*. 1977; Parrish 1993).

As noted in the previous chapter, the central part of the Atlantic had begun to open in the Mid Jurassic, and by the Mid Cretaceous the zone of opening extended northwards as far as the southern tip of Greenland, with the Labrador Sea starting to open about 95 Ma (Scotese *et al*. 1988; Rowley and Lottes 1988; Savostin *et al*. 1986). The Arctic Ocean opened by anticlockwise rotation of the North Alaska–Chukotka block away from the Canadian Arctic islands in Early to Mid Cretaceous time, between about 131 and 110 Ma (Rowley and Lottes 1988; Fig. 8.2).

The link between North America and eastern Eurasia has palaeobiogeographic significance and requires particular attention. There is no evidence of geological differences between Alaska and north-eastern Siberia; the cordilleran foldbelt of Alaska correlates with the Chukotka foldbelt. Churkin (1972) considers that the many geological features in common to the two sides of the Bering Strait indicates a close connection at least since the Palaeozoic. The westward convergence of North America on Eurasia due to the opening of the North Atlantic must, however, be indicated by compressional features somewhere. Although the Verkhoyansk Mountains have been interpreted as a plate boundary, Churkin disputes this, favouring instead the Cherskiy orogenic belt on the south-western flank of the Kolyma Massif (Fig. 8.2). A combination of stratigraphic, stuctural, and petrological features here suggest a convergent plate boundary, with the suturing having been accomplished mainly in the Early Cretaceous (cf. Rowley and Lottes 1988).

Fig. 8.2 Major tectonic features of the Arctic. Simplified from Churkin (1972).

With regard to the opening of the Indian Ocean, Antarctica–Australia began to separate from India in the Valanginian, at the same time as the South Atlantic began to open, with Madagascar–India continuing to separate from Africa. Late in the Early Cretaceous, at about 105 Ma, this latter spreading system ceased, so that Africa and India became a

single plate, and Madagascar had reached its present position relative
to Africa. The next major phase of spreading saw India separate from
Madagascar–Africa; this probably started a little before 90 Ma (Veevers
et al. 1980). By the Late Cretaceous it had travelled northwards towards
Eurasia at a rapid rate (Fig. 8.1).

The Caribbean–Tethyan zone

According to the plate-kinematic reconstructions of Pindell *et al*. (1988)
North and South America underwent divergence between the Mid Jurassic
and Late Cretaceous (Campanian). What can be called the proto-Caribbean
was formed in the Early Cretaceous as a consequence of this continental
separation. The insertion of the Farallon Plate between North and South
America in the Mid Cretaceous resulted in the north-eastward advance of
the Greater Antilles Arc and subduction of the proto-Caribbean ocean
crust (Ross and Scotese 1988; Fig. 8.3). From the Campanian to the present
relative motion between North and South America has been minimal.

The plate-kinematic and geological evolution of the Tethys belt from the
Atlantic to the Pamirs in south Central Asia has been reviewed respectively
by Savostin *et al*. (1986) and Dercourt *et al*. (1986). According to Dercourt
et al., there was a 30° anticlockwise rotation of the microplate Apulia
(including present-day Italy) with respect to Africa between about 130

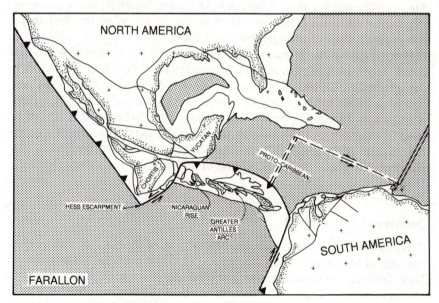

Fig. 8.3 Tectonic reconstruction for the Caribbean region in the Mid Campanian. After Pindell *et al*. (1988).

Late Mesozoic

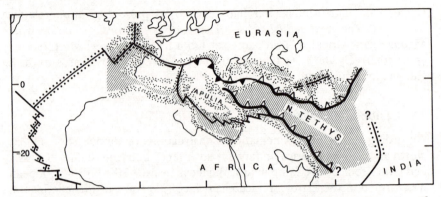

Fig. 8.4 Simplified plate tectonic scheme for the region between Eurasia and Africa at the Santonian–Campanian boundary. After Dercourt *et al.* (1986).

and 80 Ma (Fig. 8.4). Before and after these times Apulia was rigidly linked to Africa. The rotation involved a collision with Eurasia and separation from Africa. This separation resulted in the formation of a sea that includes the present-day eastern Mediterranean. Iberia was separated from Africa by a deep ocean strait throughout the Cretaceous. A drastic change took place in the interaction of the North America–South America–Africa plate system in the Late Cretaceous, between 80 and 65 Ma, with the relative motion between Africa and North America changing. An important consequence of this was the initiation of the first major compressive phase between Africa and Iberia (Savostin *et al.* 1986).

Further east in the Tethyan zone, the progressive accretion of South-east Asian terranes from Gondwana fragments has been fully discussed in the previous chapter. A reconstruction of Australian Gondwana and South-east Asia for the Early Cretaceous, by Audley-Charles *et al.* (1988) is shown in Fig. 8.5. South-western Borneo is believed to have rifted from north-eastern Indochina in the Late Cretaceous during the opening of the proto-South China Sea. It may well have formed a single tectonic block with Indochina in pre-Cretaceous times (Metcalfe 1988)

Isolation of New Zealand

Although magnetic anomalies indicate 82 Ma as the date of initiation of opening of the Tasman Sea (Weissel *et al.* 1977) it is likely that isolation of New Zealand by crustal stretching and submergence took place somewhat earlier than this. Taking into account the effect of Mid Cretaceous sea-level rise, Stevens (1980) considers that 110 Ma or even slightly earlier is a better estimate for isolation by sea.

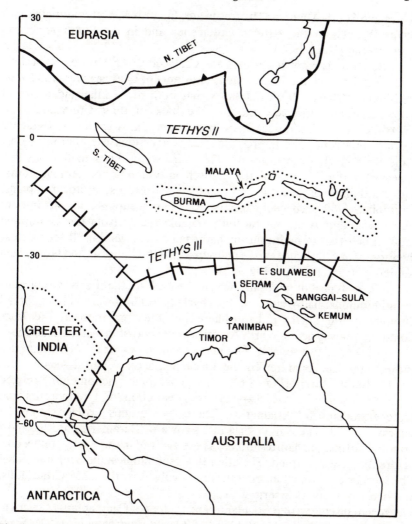

Fig. 8.5 Reconstruction of the eastern Tethys region for the Early Cretaceous (120 Ma) showing the relationship of Australia with Greater India, South-East Asia, and Eurasia. Simplified from Audley-Charles *et al.* (1988).

NON-MARINE ORGANISMS

Attention is confined in this section to vertebrates and plants, because too little is known yet about global distributions of invertebrates.

The distribution of Early Cretaceous vertebrate faunas is derived from that of the Late Jurassic and is hence vicariant. Two crocodile genera, together with the freshwater actinistian fish *Mawsonia*, are present in both

Africa and South America. The freshwater lungfish *Neoceratodus africanus* occurs throughout the African Cretaceous and in the South American Cenomanian (Rage 1988).

Much more is known about Late Cretaceous vertebrates, especially dinosaurs. Because the Late Jurassic dinosaurs of Laurasia and Gondwana are similar (Charig 1973) it follows that both areas inherited a similar fauna from the Jurassic, and that the bulk of their Cretaceous faunas could be merely descendents of common Jurassic stocks. Evidence bearing on Cretaceous intercontinental connections can therefore only come from newly evolved forms. There are seven dinosaur families that appeared in the Cretaceous, the saurischian tyrannosaurs, dromaeosaurs, and ornithomimids and the ornithischian hadrosaurs, protoceratopians, ceratopians, and pachycephalosaurs. All these groups are known in both Euramerica and Asia, but the only reliable report from the Gondwana continents is that of hadrosaurine hadrosaurs in Argentina. It seems clear therefore that a Tethyan marine belt separated the Laurasia and Gondwana continents (Cox 1974 and Fig. 8.6).

The new types of dinosaur which evolved in the Early Cretaceous (ornithomimids, hadrosaurs, and pachycephalosaurs) were able to spread throughout the northern hemisphere but the formation of the Mid-Continent seaway across North America at the end of the Early Cretaceous (Albian) created two separate landmasses, called by Cox (1974) Asiamerica and Euramerica, separated by the Turgai Sea across Central Asia but not by the Bering Strait (Fig. 8.6). This geographic phenomenon, related to eustatic sea-level rise, explains the restriction of tyrannosaurs and protoceratopians to Asiamerica. These two groups evolved after the seaway formed. The only hadrosaurs known in Euramerica are the most primitive forms, the hadrosaurines. These evolved in the Early Cretaceous, before the seaway formed. The other three subfamilies, the Saurolophinae, Cheneosaurinae, and Lambeosaurinae, evolved in the Late Cretaceous, and occur only in Asiamerica.

A recent discovery of particular interest is that of Campanian–Maastrichtian dinosaurs from the North Slope of Alaska (Parrish *et al.* 1987). Since this was near to the Cretaceous North Pole, the implication is that some dinosaurs could either tolerate three months of winter darkness or were capable of seasonal migrations across a considerable latitudinal distance. Dinosaurs of Valanginian–Albian age have also been found in southeastern Australia, a part of the world where palaeogeographic reconstructions indicate a high-latitude position at that time, about 80° S. A cool, seasonal climate is indicated by growth rings in araucarian–podocarp–gingko woods and oyxygen isotope results (Rich *et al.* 1988). The fact that dinosaurs coped with high latitudes for at least sixty five million years through the Cretaceous suggests that cold and darkness may not have been the prime factors causing their extinction, as has sometimes been maintained.

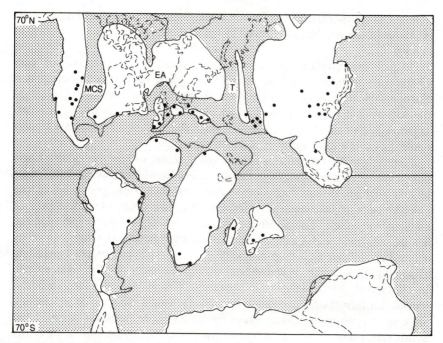

Fig. 8.6 Palaeogeographic map for the Late Cretaceous. Epicontinental seas and oceans stippled. The positions of localities containing Upper Cretaceous terrestrial vertebrates shown by black circles. EA, Euramerica; MCS, Mid-Continental Seaway; T, Turgai Sea. After Cox (1974).

The Gondwana continents had a number of distinctive vertebrate taxa missing from Laurasia. Titanosaurid dinosaurs occur in South America, Africa, India, and Madagascar. *Laplatosaurus*, a dinosaur first described from Argentina, has also been reported in India and Madagascar but, because of the poor quality of the fossils, its occurrence there is questionable. Freshwater lungfish in South America, Madagascar, and Australia are closely related to each other (Rage 1988). Madtsoiine boid snakes (Rage 1988) and trematochampsid crocodiles (Buffetaut 1985*a*) are recorded from South America, Africa, and Madagscar.

Turning to the plant record, the traditional approach has been confined to macroscopic material, but increasing attention is paid in modern studies to spores and pollen.

For the Early Cretaceous a belt approximately 30° wide on either side of the palaeoequator can be distinguished, characterized by the matoniaceous fern *Weichselia* (Barnard 1973). Its northern limit coincides closely with the boundary of Vakhrameev's (1991) Siberian Region. In some contrast to Late Jurassic Siberian floras these have abundant species of ferns, ginkgos, and true cycads. However, greatest interest in the Cretaceous is in the angiosperms. A significant advance in the understanding of their early evolution has been achieved in recent years by palynological studies (Doyle 1977).

Critical examination has convincingly refuted older claims of pre-Cretaceous angiosperms and studies of pollen in North America and Europe indicates an appearance in the Barremian and rapid diversification thereafter, such that by Cenomanian times they were already established as the dominant land plants. Study of angiosperm leaves indicates a similar pattern of radiation to that deduced from pollen.

Brenner (1976) recognized four palynofloristic provinces for the Early Cretaceous, from his studies in the Americas, based largely on differences in non-angiospermous dominants. Doyle extends these provinces to the Old World and takes due note of the differing times of appearance of angiosperm pollen. The South Laurasian Province, embracing the United States, Europe, and Central Asia, is characterized by a high diversity of conifer pollen, including many bisaccates, and fern spores, including abundant Schizaceae; angiosperm pollen appears in the Barremian. In the North Laurasian Province (arctic North America and Siberia) schizacean fern spores are much less diverse and both conifer pollen and *Sphagnum*-type moss spores proportionately more common than in the South Laurasian Province. Angiosperm pollen does not appear until the Cenomanian. As the South Gondwana Province (Australia, South Africa, Patagonia) has a first angiosperm appearance in the Mid to Late Albian, it is evident that there was a significant lag in the migration of angiosperms to higher latitudes, presumably because of the time required to adapt to cooler climates.

The North Gondwana Province embraces South America excluding Patagonia, Africa excluding South Africa, and Israel. In the coastal basins of Brazil, the earliest reticulate tricolpate angiosperm pollen occurs below the well-known salt deposits, which in turn are overlain by topmost Aptian–Lower Albian marine deposits. As the age of these tricolpates (a group that includes the vast majority of modern angiosperms) is older than the first well-dated tricolpates in Europe (Lower–Middle Albian) it is likely that they evolved in the North Gondwana Province and subsequently migrated to the South Laurasia Province. Brenner (1976) points out that the association with salt and the dominance of *Classopolis*, known to be produced by xeromorphic conifers, together with the rarity of fern spores, suggests that environmental conditions were rather hot and at least seasonally dry. This can be held to support the notions of both Stebbins (1974) and Axelrod (1970), who considered that the first angiosperms were likely to have been weedy, pioneering shrubs of disturbed habitats in seasonally dry tropical or subtropical climates. Doyle (1977) believes that the factor which triggered the radiation was probably the evolution of their key biological innovations (efficient reproduction, capacity for symbiotic interactions, vegetative flexibility) rather than large-scale changes in the physical or biotic environment.

Another attempt to determine palynofloristic provinces has been made by Srivastava (1978). Whereas only three provinces are recognized in the Neocomian (pre-Aptian Cretaceous) and two in the Albian there are five well-defined provinces in the Late Cretaceous. Although Srivastava argues that the greater amount of provinciality in the Late Cretaceous is due to the greater separation of landmasses at a time of high sea level, this is not generally evident from his phytogeographic map (Fig. 8.7). The most obvious differences apparently relate to latitude and hence probably climate. Thus there are two high latitude provinces, the *Aquilapollenites* in the north and the *Nothofagus* in the south, which extend broadly across Laurasia and Gondwana. The same is true of the low latitude *Galeacornea–Constantinisporis* and northern intermediate latitude *Normapolles* provinces. Only the Indian peninsular province was isolated from the others by ocean, with the microfloral assemblages there probably being linked with Africa (Srivastava 1978).

A multivariate statistical approach to Cretaceous phytogeographic mapping has recently been undertaken by Spicer *et al.* (1993). As with the Early Mesozoic, plant productivity was evidently concentrated in middle and high latitudes. Polar cool temperate rain forests in coastal areas, where winter temperatures were ameliorated by the proximity of the ocean, were conifer-dominated and deciduous. In more continental high latitudes, winter temperature probably fell well below freezing, thermally depressing metabolic rates and allowing some plants to retain their leaves the year round. At mid latitudes open canopy woodlands and forests were dominated by a mixture of microphyllous conifers, moderately xeromorphic ferns, cycadophytes, pteridosperms, and sphenophytes. Large angiosperm trees were comparatively rare. Although the evidence is meagre, low latitude vegetation tended to be xeromorphic and only patchily forested.

LAND CONNECTIONS BETWEEN CONTINENTS

Vertebrates are likely to provide a better guide to intercontinental connections than plants, especially pollen and spores, because of their more limited capacity for dispersal. Consequently plant distributions will not be considered in this context. The essential problem posed, at least for the Late Cretaceous, at a time of high continental dispersal and high sea level, is how did some land organisms get from one continent to another. What migrations took place, across Tethys, between the northern and southern continents (Hallam 1981*b*)? With regard to east–west connections, the

most important involve Africa and South America, Asia and North
America, and South America–Antarctica–Australia. The last of these is
more appropriately dealt with in Chapter 9.

Trans-Tethyan links

There are a number of indications of migrations of non-marine vertebrates
between North and South America in latest Cretaceous (Campanian–
Maastrichtian) and Palaeocene times (Bonaparte 1984; Rage 1988; Gayet
et al. 1992). In some cases stratigraphic data allow a determination of
polarity. Thus it can be inferred that hadrosaur and ceratopian dinosaurs
and perhaps condylarth mammals migrated from north to south. Animals
that moved in the reverse direction in the Maastrichtian include lepidosteid

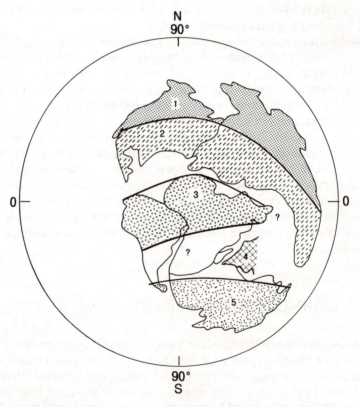

Fig. 8.7 Late Cretaceous (pre-Late Maastrichtian) palynological provinces. 1,
Aquilapollenites Province; 2, *Normapolles* province; 3, *Galeacornea–Con-
stantinisporis* province; 4, India peninsula province; 5, *Nothofagus* province.
After Srivastava (1978).

fish, teiid lizards, anilid and boid snakes, and titanosaurid dinosaurs. Migration from South to North America in the latest Cretaceous or Palaeocene included osteoglossid and ariid fish, iguanid lizards, and some snakes and mammals. Didelphid mammals also intermigrated between the subcontinents, but the direction of migration is not known. None of the groups involved appear in either North or South America prior to the Campanian. Therefore exchanges probably began then and increased in the Maastrichtian and Palaeocene.

Some widely distributed terrestrial groups in the Laurasian and Gondwanan continents failed to cross. This failure could signify a discontinuous land route necessitating island hopping, but some of the organisms in question were freshwater fish, which require a strictly non-marine route for dispersal. Others, such as the titanosaurids, were very large, and hence would have found island hopping rather difficult. Accordingly it is likely that other environmental factors restricted the migration of some groups. The most probable migration route was the Greater Antilles and Aves Ridge, which consisted of a magmatic arc submitted to uplift and deformation between North and South America at that time (Fig. 8.3). Part of the arc began to collide with both subcontinents in the Campanian, the time when the faunal exchanges began. The other migration possibility, the volcanic arc to the south-west of the Caribbean Plate, was probably highly discontinuous and did not constitute a land bridge (Pindell *et al.* 1988).

With regard to a possible land connection between Europe and Africa, Rage (1988) argues that, if there were such a direct connection, there should have been a greater similarity of faunas. Such interchanges as there were seem to have been rather rare, despite the Apulian Plate providing a series of emergent lands between Europe and Africa (Dercourt *et al.* 1986). The pattern as recognized is consistent with the African endemism which is more apparent in the Palaeogene.

The case of the Indian subcontinent is especially interesting. All plate tectonic reconstructions, for example Fig. 8.1, have shown India isolated by ocean in the Late Cretaceous, and the time of collision with Asia has been generally accepted as Palaeogene. However, the discovery of pelobatid frogs in sediments within the Deccan Traps basalts in Nagpur, dated as Maastrichtian, indicates terrestrial faunal exchange with Laurasia (Sahni *et al.* 1982; Sahni 1984). There is no evidence from these sediments, the Takli Formation, of any endemism, as would be expected from a long period of isolation by ocean. This inference has been supported by the discovery of palaeoryctid mammals in intertrappean beds of the Naskal Formation of Asifabad (Prasad and Sahni 1988). These are typical Laurasian mammals known to occur both in the Upper Cretaceous of Mongolia and North America. More recently, the Naskal Formation in Andhra Pradesh has yielded a discoglossid frog, another vertebrate of Laurasian affinities (Prasad and Rage 1991). Amphibians cannot survive salt water and therefore a continuous land connection is required.

These discoveries have required a revision of the plate tectonic story for India (Jaeger *et al.* 1989). The time at which a major change in relative motion between India and Asia seemed to have taken place was originally thought to reflect the onset of collision, about 50 Ma. Joint analyses of both primary and secondary components of magnetization in Upper Palaeocene sediments from the northernmost margins of India also led Besse *et al.* (1984) to propose a collision age of 50 Ma. However, the faunal data indicate that the collision was much earlier. It implies a north–south intracontinental shortening of about 4000 km rather than 2000 km, as formerly believed. This in turn implies either a large extension of the northern Indian margin or large scale strike-slip *extrusion*, especially in Indochina, along the Red River Fault. Fig. 8.8 indicates a compromise between these scenarios, with Greater India considered to have extended up to 1000 km north of the Main Boundary Thrust, and with about 1500 km strike-slip-related shortening. The timing of collision is close to the age of eruption of the Deccan Traps, but this could be coincidental, since these are generally related to extrusion of the Reunion hotspot.

Africa–South America

Buffetaut and Rage (1993) have reviewed the evidence from fossil amphibians and reptiles bearing on this connection. With regard to the Early Cretaceous, a land continuity has long been required to account for the extremely close resemblance of non-marine ostracod faunas of Brazil and tropical Africa (Krömmelbein 1965a, 1965b, 1971). This is confirmed by the reptiles found in Aptian deposits in north-eastern Brazil (the Santana Formation) and Niger. There are two crocodile genera in common in the non-marine deposits. The markedly terrestrial adaptations of *Araripesuchus* make it very unlikely that it could have crossed marine barriers. There is also a close resemblance among the turtles, but very little is known about the Early Cretaceous dinosaurs.

By the Late Cretaceous, as noted earlier, the South Atlantic had become well established, with all connection being severed between 95 and 80 Ma. There are two transverse ridges formed by rift volcanism, the Ceará and Sierra Leone Rises in the north and the Rio Grande Rise–Walvis Ridge to the south, that could possibly have provided at least island stepping stones for some time after this. The precise time when the Rio Grande–Walvis Ridge was severed as a continuous land connection is unknown, but Parrish (1993) considers it to be probably mid Early Cretaceous. The data for the more northerly ridge are difficult to evaluate.

The pipids are freshwater frogs confined today to Africa south of the Sahara and tropical South America and the African Plate, including Israel.

The oldest known fossils are from the Lower Cretaceous of Israel. Upper Cretaceous fossils are known from Argentina and Niger. Two alternative palaeobiogeographic models are proposed by Buffetaut and Rage: (1) an early dispersal followed by a vicariant event as the South Atlantic opened; (2) a late dispersal, to account for the close relationship of species of *Silurana* from the Brazilian Palaeocene and Recent African forms. They speculate that a Walvis–Rio Grande filter bridge could have been utilized in the Late Cretaceous or Palaeocene, but this does not accord with the geological evidence reviewed by Parrish. Furthermore, the absence of hadrosaurs from Africa implies the lack of a Late Cretaceous land connection with South America.

Among the reptiles, madtsoiid snakes are known from the Campanian and Maastrichtian of South America and the Upper Cretaceous of Africa

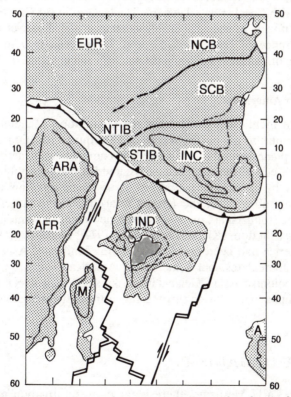

Fig. 8.8 Possible palaeogeographic reconstruction of India and continents bordering Tethys and Indian oceans at Cretaceous/Tertiary boundary time. EUR, Eurasia; NCB, North China block; SCB, South China block; NTIB, northern Tibet; STIB, southern Tibet; INC, Indochina; M, Madagascar; IND, India; AFR, Africa; ARA, Arabia; A, Australia. After Jaeger *et al.* (1989).

and Madagascar. The oldest snakes known are Albian in age and the oldest madtsoiids Coniacian at the earliest, so the group may have arisen before or after the South Atlantic opening. Then therefore, they cannot be used to establish whether vicariance or dispersal is required to explain the distribution.

It is believed that subsequently an African group must have crossed the ocean barrier by Late Cretaceous–Early Palaeocene times. This is based on the *Madtsoia-Gigantophis* group occurring in the African Upper Cretaceous whereas the oldest representatives in South America are of Palaeocene age. Since there are rich herpetofaunas in the South American Upper Cretaceous their absence is probably significant. Once more a continuous Walvis–Rio Grande land route is invoked by Buffetaut and Rage, an idea which is subject to the same difficulty alluded to in the previous paragraph. One family of non-marine mesosuchian crocodiles, the Trematochampsidae, occurs in the Upper Cretaceous of Africa and South America, with closely related genera. The lack of Lower Cretaceous fossils makes it difficult to choose betwen vicariance and dispersal hypotheses. The occurrence of both *Madtsoia* and *Trematochampsa* in the Upper Cretaceous of Madagascar poses the same problem (Rage 1988).

Asia–North America

It has been mentioned earlier that the distribution of Late Cretaceous dinosaur faunas in the northern continents requires isolation by two mid-continent epicontinental seas but free land communication across Beringia. Towards the end of the Cretaceous, Asiamerica was probably subdivided, presumably by an epicontinental sea across Beringia. This is suggested by the likely absence of the latest Cretaceous North American family Ceratopidae from Asia (Cox 1974). Cox's view receives some support from mammal distributions (Kielan-Jaworowska 1974). By the latest Cretaceous, North America had been invaded by several families of Asian placentals and multituberculates, but some North American forms were evidently unable to colonize Asia. Kielan-Jaworoska suggests a sweepstakes route across Beringia, with marine currents favouring eastward but not westward crossings.

MARINE ORGANISMS

As for the Early Mesozoic, there is far more information available for marine than non-marine fossils. Although some of the most biogeo-graphically illuminating organisms belong to the latter group, study of the distribution of marine fossils has yielded much valuable information about the oceans and climate.

Latitude-related realms

Similar to the situation in the Triassic and Jurassic a low-latitude Tethyan Realm can be discerned, characterized by abundant carbonates and reef-forming organisms (Fig. 8.9). Unlike the two earlier Mesozoic periods rudist bivalves dominate over corals and their marked provincialism forms the basis for identifying finer biogeographic units. The bivalve faunas also include diverse trigoniids and lucinids, and a high diversity of epifauna and borers, together with a high percentage of thick shelled, ornate forms that had a shallow-water infaunal habitat (Kauffman 1973). The most characteristic Tethyan gastropods belong to the Nerinacea and Actaeonellidae. The most diverse superfamily, the nerinaceans, suffered a precipitous decline after the Cenomanian, but the actaeonellids maintained a high diversity until the Maastrichtian (Sohl 1987).

Another important Tethyan group is the larger benthic foraminifers, with such genera as *Orbitolina* and *Orbitoides* widespread throughout the region in the Mid and Late Cretaceous. The Globigerinacea, a planktonic group, became important in the Late Cretaceous and are largely confined to lower latitudes, though not completely restricted to the Tethyan Realm (Dilley 1971).

Both ammonites and belemnites clearly indicate a Boreal Realm for the Early Cretaceous but the Boreal–Tethyan distinction disappeared in the Late Cretaceous (Casey and Rawson 1973; Matsumoto 1973; Doyle 1992). The belemnite distributions are particularly interesting (Stevens 1973*b*; Doyle 1992). In Berriasian–Hauterivian times the pattern compared with that for the Jurassic, with the Cylindroteuthidae being the dominant Boreal family. The Tethyan belemnopsid *Hibolithes* and some duvaliids reached as far north as East Greenland during short-lived episodes. In the European Tethys *Hibolithes* and duvaliids persisted and diversified during this time, extending as far as Indonesia and Madagascar. What Doyle calls the Gondwanan Tethyan Province was dominated by *Hibolithes* and *Belemnopsis* (the latter genus had disappeared from Europe in the Oxfordian).

From Barremian times onwards widespread changes took place. In the northern hemisphere, *Acroteuthis*-dominated assemblages declined and retreated to the Arctic Province of the Boreal Realm. In the Aptian, *Neohibolites* and *Parahibolites* replaced *Hibolithes* and Duvaliidae. *Neohibolites* is the last surviving genus of a truly Tethyan stock; it continued to the Cenomanian. In Aptian–Cenomanian times an almost bipolar distribution developed. The Belemnitellidae were the first Belemnopseina to invade boreal habitats after *Neohibolites*, and replaced Belemnitida in northern Europe. Dimitobelidae appeared in the Aptian, and were restricted to seas surrounding the old Gondwana continents, southern Africa, southern India, Australasia, Antarctica, and South America. By

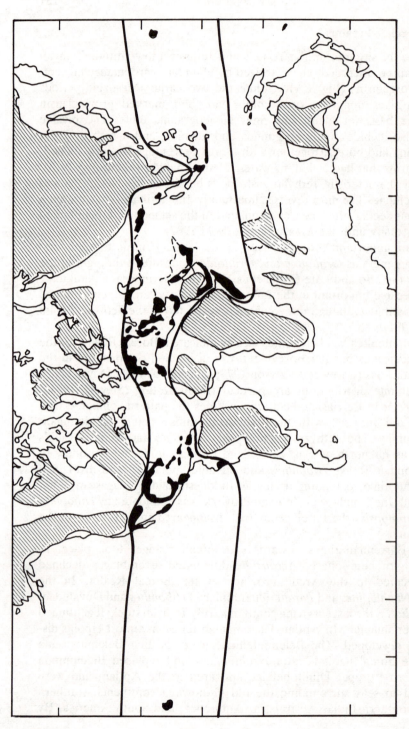

Fig. 8.9 Cretaceous distribution of algal-coral, coral-rudist, or rudist–dominated organic framework (reef) building. Heavy solid lines represent northern and southern limits of framework construction. Areas in black represent known framework fossil assemblages. Diagonal pattern represents land areas never covered by Cretaceous seas. Simplified from Sohl (1987).

the Campanian–Maastrichtian the family had become restricted to an Antarctic–New Zealand 'Weddelian' Province. Concerning the breakdown of Boreal–Tethyan provinciality, Doyle (1992) considers that the physical effects of palaeogeography and palaeoceanography were of paramount importance, as the continents separated, and climatic effects subordinate.

As with the Jurassic the other molluscan faunas reveal no distinctive Boreal Realm. Kauffman (1973) distinguishes North and South Temperate Realms flanking the Tethyan Realm, each subdivided into a number of provinces. For the important bivalve group of inoceramids, Dhondt (1992) fails to recognize a distinction between Tethyan and non-Tethyan forms. Sohl (1987) disagrees with Kauffman that diversity is always lower in the Temperate than the Tethyan Realm. In particular, the greater abundance and diversity of Late Cretaceous neogastropods in the Temperate Realm, in contrast to the situation today, supports the idea that the group arose in higher latitudes.

With regard to planktonic organisms, Mutterlose (1992a) considers that it is possible to differentiate between Boreal and Tethyan Realms among Berriasian–Barremian coccoliths, but subsequently there was a global homogenization of calcareous nannofloras.

The Cretaceous marks the earliest time in the Mesozoic that distinctive austral biotas became well established, both on land and in the sea (Crame 1992b). Much of the provinciality is centred on Australasia (Stevens 1980). This has already been illustrated with reference to the Dimitobelidae. However, there is nothing to compare with the rich ammonite and belemnite boreal biotas. Thus, in Kauffman's (1973) Austral Province of the South Temperate Realm, there is only one endemic bivalve genus, *Maccoyella*.

The effect of Atlantic opening

As the North Atlantic Ocean progressively opened up and widened tropical neritic faunas on the two sides should have diverged (higher latitude faunas could have migrated with ease because no oceanic gap yet existed between Greenland and northern Europe). This is in fact the case is clearly demonstrated for corals and rudists by Coates (1973). In Aptian–Albian times the Caribbean region appears as a distinct province for the first time. In general, Caribbean rudists show a consistently higher percentage of endemic genera than corals but both groups follow similar trends. The highest diversity and level of endemism is in the Maastrichtian. Other bivalves show a parallel trend (Fig. 8.10).

Similarly, endemic actaeonellid species appear in the Caribbean region in the Campanian (Sohl 1987) and generic links between low-latitude brachiopod faunas of Europe and North America became much reduced in the Late Cretaceous (Sandy 1991). Comparable changes are also recorded

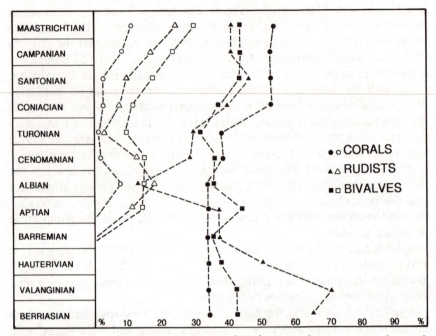

Fig. 8.10 Graphs showing percentage of Tethyan endemic genera for each stage in the Cretaceous for the Caribbean Province (open symbols) and the Mediterranean Province (black symbols). After Coates (1973).

for another Tethyan group, the larger benthic foraminifers. Dilley (1973) infers that east–west crossing was easy in the Early Cretaceous but that a barrier was established in the Mid Cretaceous, with the entire group disappearing from Central America at this time. In the latest Cretaceous it reappeared, but the evidence is strongly indicative of the difficulty in making the east–west oceanic crossing.

With regard to Temperate Realm or more cosmopolitan taxa, one needs to bear in mind that parochial taxonomy has almost certainly led to an underestimate of the similarities of North American and European species (Dhondt 1992 and C.F. Koch, personal communication). The presence of 'typically Tethyan' brachiopods in the Valanginian of East Greenland, most notably *Pygope*, has been interpreted as an indication of an early 'Gulf Stream' (Ager and Walley 1977). Pygopids have been interpreted as the inhabitants of deeper Tethyan seas, so they could indicate cool rather than warm water. The associated 'Boreal' fauna of large terebratulids and ribbed rhynchonellids could signify a shallower-water high energy fauna (Sandy 1991).

The establishment of a continuous marine connection between the northern and southern proto-Atlantic by the Albian is clearly demonstrated by ammonite distributions (Kennedy and Cooper 1975; Förster 1978). The occurrence of deep-water morphotypes of planktonic foraminifers confirms the presence of deep oceanic circulation, at intermediate water depths, between Brazil and low-latitude western Tethyan regions since the Late Albian. Good oceanographic connections between the North and South Atlantic since the Early Turonian are indicated by large-keeled taxa such as *Dicarinella* and *Marginotruncana* (Koutsoukos 1992).

A cautionary note is struck by Smith (1992) in his analytical study of Cenomanian echinoids. He considers that, whereas for Recent biotas biogeography can be said to have entered Ball's (1976) analytic phase, Cretaceous biogeography is still in his narrative phase and based on phenetic rather than cladistic taxonomy. Smith's cladistic vicariance analysis shows that Texas and South Atlantic (Brazil, Angola) are either part of a near-cosmopolitan fauna or are the sister group of eastern Atlantic rather than North African faunas. Echinoid distribution patterns in the Cenomanian conform to a mosaic model rather than a model of discrete provinces.

Trans-Pacific migration

The Indo-Pacific region had some distinctive characteristics in Cretaceous times. Thus in the Late Cretaceous the ammonite *Kossmaticeras* and related genera occur commonly in areas surrounding the Pacific and Indian Oceans, where phylloceratids, tetragonitids, and desmoceratids also had their domain (Matsumoto 1973). For Cretaceous bivalves Kauffmann (1973) distinguishes a North Pacific Province and Dhondt (1992) points out that this region had inoceramid faunas different from the rest of the world, a rather interesting fact bearing in mind the generally cosmopolitan nature of the group.

More intriguing still, however, is the occurrence in the north-western Pacific of seamounts capped by Barremian–Albian rudist reefs, clearly indicating that they are drowned atolls. Palaeomagnetic results indicate that they grew approximately 30° south of their present location, in the South Pacific Polynesian region (Winterer 1991). These reefs are contemporary with the 'Urgonian' of western Europe and Middle Eastern rudist reefs, while in the Caribbean–Gulf of Mexico region their greatest development is in the Albian and Cenomanian.

The hippuritid *Torreites* is absent from the well studied and faunally rich European deposits and was formerly considered endemic to the Caribbean, but it has now been found in the United Arab Republics and Oman. Some other Campanian and Maastrichtian rudists and gastropods show a similar pattern of distribution, occurring in the eastern Tethys and Caribbean but not in the western Tethys (Skelton 1988). Skelton infers that the shallow-water plateaus and seamounts formed by mid-plate volcanism and swells in the Pacific (Winterer 1991) have served as staging posts for trans-Pacific dispersal of larvae. Since some of the taxa in question occur in slightly older strata in the Caribbean, the data are consistent with an east to west dispersal. In the earlier Cretaceous, larvae could only get as far as the eastern or central Pacific. This is because volcanic swells were formed further east and have since migrated to the north-west, the direction of sea-floor spreading. The western Pacific was probably floored by cold, deep ocean floor, forming a barrier comparable to the present eastern Pacific. Subsequently communication again failed as the swells subsided. Thus Palaeogene larger benthic foraminifers in reefal material of the central Pacific are exclusively of Old World affinities.

Oceanographic and climatic effects

Episodes of sea-level rise and concomitant marine transgressions over the continents were associated with diversity increases and the reverse process with diversity decreases, at least among the bivalves (Kauffmann 1973). Pulses of sea-level highstands in the Neocomian enabled Tethyan nannofloras and faunas (most notably ammonites and belemnites) to spread into north-western Europe (Mutterlose 1992b). The major eustatic rise commencing in the Aptian led by Mid Cretaceous times to a tendency towards global homogenization of many marine biota, as exemplified for example by coccoliths (Mutterlose 1992a) and ammonites (Riccardi 1991).

Whereas the general pattern associated with marine transgression is a reduction of endemism, the Western Interior or Mid Continent seaway of North America provides an exception. This seaway was created as a result of the Mid Cretaceous eustatic sea-level rise, and became the locus of endemism among a number of groups, including scaphitid and baculitid ammonites (Matsumoto 1973), bivalves (Kauffmann 1973), and benthic foraminifers (Dilley 1971). Evidently communication with the open ocean system was restricted to some degree, allowing endemic faunas to evolve.

With regard to climate, it has been well established that the terrestrial plant record indicates an appreciably more equable world than today, with temperate forests extending to the polar regions (Dettman 1989; Frakes *et al.* 1992). According to Frakes *et al.* North American palaeobotanical work suggests a temperature rise from the Albian to a Santonian optimum followed by a decline in the Maastrichtian. Data from marine organisms supports that from terrestrial plants in indicating greater equability. Thus factor and principal components analyses of Middle Cretaceous coccolith distributions indicate that cool-water assemblages are largely restricted to high southern palaeolatitudes (south of 50°S), with only a small component of such high-latitude species in England and France. A wide tropical–subtropical zone is recognizable where nannofossil distribution is fairly homogeneous (Roth and Krumbach 1986). The biogeographic distribution of these low-latitude assemblages is largely controlled by surface water fertility. High-fertility ('upwelling') assemblages are dominant along the eastern margins of the Atlantic and oceanic assemblages in the central and western Atlantic.

The boundaries of the Tethyan Realm varied with time, generally being broadest in the Aptian–Turonian and more latitudinally restricted in the Coniacian–Maastrichtian (Sohl 1987). If temperature is the main control on the location of the boundaries, as Sohl believes, then a Mid Cretaceous rather than late Cretaceous climatic optimum is implied. The apparent lack of concordance between results from marine and terrestrial organisms is a subject clearly deserving of fuller investigation.

Global cooling during the Late Maastrichtian has been invoked to account for the time–transgressive migration of the calcareous nannofossil *Nephrolithus frequens*, which shifted to lower latitudes (Wise 1988). Other evidence of Late Maastrichtian cooling has been derived from oxygen isotope studies and North Atlantic planktonic foraminifer distributions (Olsson 1977). However, Late Maastrichtian poleward shift in biogeographic realm boundaries and the poleward migration of several keeled and non-keeled planktonic foraminifers directly conflict with evidence of cooling at this time (Huber 1992). Previous models have assumed that temperature was the primary factor controlling the distribution of this group. The general view has been that coarsely ornamented and/or keeled morphotypes such as *Pseudotextularia* and *Globotruncana* were restricted to tropical and subtropical waters, and that the occurrence of these forms in high latitudes reflects poleward expansion of the Tethyan Realm.

According to Huber, *Pseudotextularia elegans* is the only species having a delayed high-latitude occurrence that can be correlated with independent isotopic evidence for climatic warming. Synchronous migration to both polar regions testifies to the global extent of a postulated latest Maastrichtian warming trend (66.9–66.6 Ma). The problem arises how to corroborate Maastrichtian incursions of other 'warm stenothermal' taxa in southern high latitudes when there is overwhelming evidence for global cooling at the same time. In Huber's view, changes in the density structure and stability of surface waters may have played an equal if not more important role as temperature in governing the distribution of Late Cretaceous planktonic foraminifers. Deep-dwelling forms such as *Globotruncana* may have been excluded from high-latitude oceans during times when the water column was highly convective and density stratification poorly developed and/or strongly seasonal. Times of stability and enhanced vertical stratification of polar surface waters would have allowed the migration of such taxa to higher latitudes. Such transitions could have been controlled by global eustatic changes and periodic opening of an Antarctic Peninsula–South American isthmus.

9

Palaeogene

The Palaeogene or Early Tertiary, comprising the Palaeocene, Eocene, and Oligocene epochs, can be said to mark the time when the major outlines of the modern world, in terms of palaeogeography, climate, and biota, became progressively established. As in previous chapters attention will be paid to the geology before turning to organic distributions.

MAJOR GEOLOGICAL EVENTS

A number of important events involving sea-floor spreading can be established from the recognition of magnetic anomalies on the ocean floor. The North and South Atlantic continued to open, extending into a separation of North America from Eurasia as the Norwegian Sea began to open in the Palaeocene. Sea-floor spreading in the Labrador Sea and Tasman Sea ceased at this time. A rapid separation of Australia from Antarctica took place. This was originally thought to commence at anomaly 19 time (Mid Eocene) but revised magnetic anomaly identifications suggest that the initiation of sea-floor spreading was as early as anomaly 24 time (Late Cretaceous) preceded by a phase of initial rifting (Cande and Mutter 1982). The eastward translation of the Caribbean Plate relative to North and South America continued, and the Alps formed as a consequence of the Apulian 'prong' of Africa compressing southern Europe. Further east the main collision of India with Asia resulted in a major plate re-organization (Scotese *et al.* 1988). Fig. 9.1 gives a plate reconstruction for the Oligocene.

Opinions have varied on the collision chronology of India and Asia. Magnetic anomalies on the ocean floor and palaeomagnetic latitude measurements have been used in reconstructing this. The time of collision has been variously dated as approximately Late Palaeocene (Powell and Conaghan 1973) or Eocene (50 Ma – Besse *et al.* 1984; 40 Ma – Molnar and Tapponnier 1975). According to Chatterjee (1992) magmatic activity, tectonism, and stratigraphy along the Indus–Zangbo Suture provide better constraints and suggest a much earlier collision, with emplacement of ophiolites and granites indicating a Cretaceous subduction zone. Major deformation in southern Tibet is dated as Late

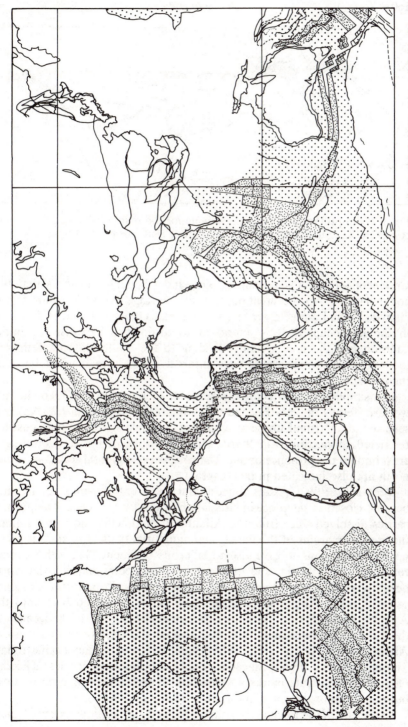

Fig. 9.1 Plate reconstruction for the Oligocene (38 Ma). Coarse stippling signifies Jurassic and Cretaceous ocean floor, fine stippling signifies Early Tertiary ocean floor. Simplified from Scotese *et al.* (1988).

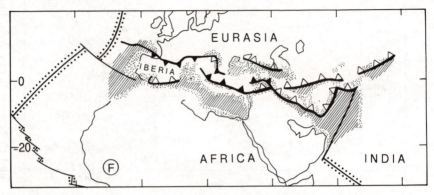

Fig. 9.2 Simplified plate tectonic scheme for the region between Eurasia and Africa in the Late Eocene (Priabonian). After Dercourt *et al.* (1986).

Cretaceous–Early Palaeocene in age, and Chatterjee believes that this could have led to the creation of a filter bridge in the Kashmir region allowing some Laurasian terrestrial vertebrates to enter India, as discussed in the last chapter. It is important to distinguish between the initial collision and the final welding, leading to the obliteration of Tethys in the region. A considerable time interval could have elapsed between these events.

The simplified plate-tectonic scheme of Dercourt *et al.* (1986) for the western Tethys in Late Eocene times is presented in Fig. 9.2. Africa is seen separated from Europe by a band of deep ocean, except for a possible restricted connection with Iberia where the zone of thinned continental crust implies the presence of sea. The zones of continental collision further north are also indicated in this diagram.

Deep-sea drilling results suggest that the Central American Isthmus became closed to deep-ocean circulation in the Late Eocene. Deep water carries dissolved silica from the Atlantic to the Pacific and upwells along the eastern margins of the Pacific and equatorial belts. The silica is taken up by radiolarians to give rise to siliceous sediment. The silica cannot return to the Atlantic, and most of the silica today entering the world ocean will ultimately remain in Pacific sediments. Prior to the existence of the Central American Isthmus, deep water circulated back into the Atlantic and supported a silica-rich equatorial belt in both oceans. In the Late Cretaceous and Early Palaeogene the Caribbean and western Atlantic silica accumulations were equal to those in the Pacific today. The change from highly siliceous Middle Eocene to much less siliceous Oligocene sediments marks a restriction to free inter-ocean circulation, but the depth of the sill is not known (Donnelly 1985).

At the end of the Eocene the isthmus formed a continuous morphological and structural unit but was not yet emergent. The tectonic reconstruction for the Caribbean by Pindell *et al.* (1988; Fig. 9.3) indicates that the Central American Isthmus was substantially established as continental crust by the Early Oligocene, as a result of volcanic activity associated with the subduction zone directly to the south-west. This does not necessarily imply the existence of a continuous land corridor between North and South America; there were marine gaps, the largest one being in the Panama region.

With regard to sea level, following the substantial fall at the end of the Cretaceous there were subsequent rises in the Palaeocene and, more considerably, in the Eocene, though nothing to compare with the Mid–Late Cretaceous rise, and the continents remained substantially emergent. Among the sea-level falls, by far the biggest took place in the Mid Oligocene, possibly up to as much as 100 m (Hallam 1992 and Fig. 4.1).

Climatic changes were even more striking. Based on data from the changing distributions of marine plankton and terrestrial plants, and oxygen isotope results from foraminifers, increasing temperatures to an Early Eocene high were followed by a fall, accelerating across the Eocene–Oligocene boundary. This latter change marks the passage from a Warm to a Cool Mode in the Phanerozoic scheme of Frakes *et al.*

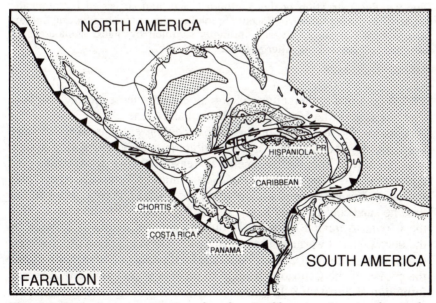

Fig. 9.3 Tectonic reconstruction for the Caribbean region in the Early Oligocene. After Pindell *et al.* (1988).

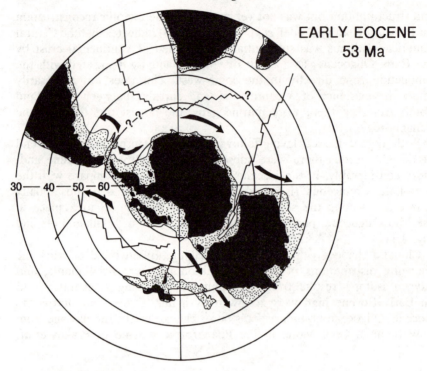

EARLY EOCENE
53 Ma

Fig. 9.4 Reconstruction of the Southern Ocean and suggested bottom-water circulation during the Early Eocene. Spreading ridges and connecting fracture zones shown at jagged lines. Continents signified by black, shelf areas by stippled ornamentation. After Kennett (1980).

1992 (Fig. 4.1) and thereafter global temperatures never returned to those characteristic of the Mesozoic and Early Palaeogene.

A major forcing factor in this global cooling event is likely to have been the development of a circum-Antarctic oceanic circulation system associated with the establishment of seaways between Australia and Antarctica and Patagonia and the West Antarctic Peninsula (the Drake Passage). This latter event took place in the Early Oligocene. Prior to this time no such current could have existed (Fig. 9.4). With the formation of the Circum-Antarctic Current, Antarctica became thermally isolated by the decoupling of warmer subtropical gyres. These migrated progressively northwards as the cold Circum-Antarctic Current increased in width during the course of the Cenozoic, leading to increased glaciation and ice-sheet formation (Kennett 1980).

DISTRIBUTION OF TERRESTRIAL ORGANISMS IN RELATION TO PALAEOGEOGRAPHY AND CLIMATE

Whereas in previous chapters vertebrates and plants have for the most part been treated separately, for the Palaeogene both groups can contribute information on a number of topics. Generally, vertebrates are more informative for continental relationships, plants for climate.

Laurasian continental connections

Of 60 Lower Eocene (Sparnacian) mammal genera known from Europe, 34 also occurred in North America but only two genera, *Pachyaena* and *Coryphodon*, are known with reasonable probability to have occurred in both Europe and Asia. This clearly indicates a direct land connection across the North Atlantic region, with Asia being isolated from Europe by a seaway across the West Siberian lowlands linking the Arctic Ocean with Tethys via the Turgai Strait, north-east of the Caspian Sea. In the Mid Eocene the link with North America was severed by sea-floor spreading in the Norwegian Sea (McKenna 1975). McKenna's conclusions were anticipated by Kurtén (1966) and are further supported by the Early Eocene reptile fauna of the Canadian Arctic, which suggest continuity between North America and Europe (Estes and Hutchinson 1980).

Both in North America and Europe there was a major turnover of mammal faunas in the Late Eocene (Prothero 1989). Thus in North America a number of archaic groups, including titanotheres, multi-tuberculates, and various rodents declined and eventually became extinct at the end of the epoch. The extinction event in North America preceded the first appearance there of canids, felids, mustelids, tapiroids, and rhinocerotoids, with many new groups evidently having arrived via Beringia (Webb 1985). Similar patterns are observed in Europe, with the so-called 'Grande Coupure' marking the end-Eocene extinction. Mid Eocene faunas were dominated by primates, multi-tuberculates, creodonts, condylarths (archaic ungulates), pantodonts, and archaic perissodactyls and artiodactyls. After the Grande Coupure the new fauna was dominated by rodents, advanced carnivores, artiodactyls, and perissodactyls. All archaic groups, particularly creodonts, archaic ungulates, multi-tuberculates, tillodonts, and pantodonts, became extinct. According to McKenna (1983) nearly all of the new western European occurrences had obvious Asian or American affinities or both, implying immigration after a new dry land connection had become established. This was evident across Eurasia as the epicontinental sea withdrew at the start of the Oligocene, because the Atlantic Ocean at this time linked with the Arctic Ocean. A Bering land bridge connection between Asia and North America persisted, at least intermittently.

India

The Upper Palaeocene and Lower Eocene deposits of the Indian subcontinent contain some endemic mammals, for which the geographic origin cannot yet be located (Sahni *et al.* 1981). There is a diverse mammal fauna from the Middle Eocene, which has strong and indubitable affinities to the contemporary faunas of Mongolia and Kazakhstan (Sahni and Kumar 1974; Sahni *et al.* 1981). The immigration of large terrestrial mammals including hyracodontid rhinoceratoids, tapiroids, and artiodactyls south of the Indus–Zangbo suture zone clearly indicates that a major land corridor between India and Asia had become established by Mid Eocene times. Prior to this intermigration could have been achieved by filter bridges. No intercontinental exchange can be demonstrated during the Palaeogene between the African and Indian Plates.

Africa

Until recently the earliest Tertiary mammals known in Africa were of Mid Eocene age, and consisted of marine groups, cetaceans and sirenians. Proboscideans, which originated in Africa, occur in the Upper Eocene of Egypt; they share with the sirenians a condylarth ancestry. The Eocene land fauna was wholly endemic and had been apparently isolated in Africa for a long time (Coryndon and Savage 1973).

Much richer faunas come from the Lower Oligocene, most notably from Fayum (Egypt) and Libya. According to Coryndon and Savage (1973), of 31 genera, 25 are exclusively African. These latter include proboscideans, hyracoids, and embrithopods. Other taxa evolved in Africa from ancestors that crossed from Eurasia prior to the Early Oligocene. Traffic was one-way into Africa, which argues strongly against any land bridge, and chance dispersal seems more likely. The celebrated Fayum deposits have recently yielded the first marsupial known from Africa (Bown and Simons 1984). It is a didelphid, a group known from European deposits of Early Eocene to Mid Miocene age. Bown and Simons consider that, because there was no free land connection between Europe and Africa in the Early Oligocene, dispersal into Africa was probably by means of a sweepstakes route.

In the 1980s the mammal record was extended back to the Upper Palaeocene, with the discovery of several new North African localities. The Eocene ones confirm a distinct African endemism but indications exist of interchanges with North Tethyan areas in the Palaeocene (Gheerbrant 1990). The Palaeocene faunas contain palaeoryctids comparable to those in North America. European affinities are indicated by *Adapisoviculus* and *Afrodon*, both present in Europe and Africa. The similar diversity of the adapisoviculids indicates that a centre of dispersal cannot be identified. North America was probably indirectly involved through the

North America–Europe connection at this time. Thus the North American affinities probably reflect gaps in our knowledge of European faunas. Gheerbrant speculates that the Cretaceous–Tertiary boundary was the most likely time for dispersal, because of the low sea-level stand.

Gheerbrant also infers a later dispersal event. Several species in Morocco support the hypothesis of Gingerich (1986) of the African origin of some groups well developed in Laurasia in the Eocene but unknown or questionable earlier. This has been suggested for artiodactyls, perissodactyls, hyaenodontids, amomyids, and adapid primates. The oldest known amomyid species is that found in Morocco, together with two hyaenodontid species. The most probable time of dispersal is likely to have been at a low sea-level stand at the Palaeocene–Eocene boundary and the most likely migration route via the Apulian connection.

With regard to non-mammalian vertebrates, Buffetaut (1982) reports the occurrence of a ziphodont mesosuchian crocodile from the Eocene of Algeria. This group occurs in the Eocene of Europe and the Palaeogene of South America, but not in the richly fossiliferous and well-studied deposits of North America. Other vertebrate groups with this distribution include amphibians, land birds, and anteaters. Buffetaut speculates on the presence of a direct land link across the South Atlantic, for instance along the Rio Grande–Walvis Ridge, but as noted in the previous chapter there are doubts about this having been fully emergent, so as to allow an intercontinental land bridge, even back in the Late Cretaceous. However, according to Thiede (1977) the Rio Grande rise finally subsided beneath the surface of the ocean in the Late Oligocene. The Walvis Ridge was probably completely submerged by the end of the Eocene (Parrish 1993). Some sort of sweepstakes route is more plausible than a continuous land corridor, not just because of geological considerations but also the strong endemism of African and South American mammals.

South America

After the limited faunal interchanges with North America that had taken place near to the Cretaceous–Tertiary boundary, South America became isolated by sea until late in the Neogene, when free intercommunication between the two continents took place, as will be discussed in the next chapter. Certain edentates, notoungulates, and uintatheres among the mammalian fossils of North America provide evidence for a Late Palaeocene dispersal event from South America. However, North American Early Eocene rodents and primates, appearing shortly after the dispersal event, were unable to invade South America (Gingerich 1985). In the interim a substantially endemic vertebrate fauna evolved. This includes such groups as sebecosuchian crocodiles (Gasparini *et al.* 1993), boid snakes (Albino 1993), and a diverse mammal fauna, including didelphid and borhyaenid marsupials. However, there are, two interesting exceptions to this mammalian endemism.

The hystricognath–hystricomorph rodents (guinea pigs, coypus, capybaras, porcupines) are largely confined today to South America and Africa, and are known as fossils from the Lower Oligocene of those two continents. There are three familes in the Fayum deposits of Egypt and a smaller diversity in the Upper Eocene of North Africa. A diverse fauna has also been found in Bolivia and Patagonia (George 1993). Although some authorities have argued for a direct crossing between the Americas, George favours the hypothesis of Hoffstetter (1972) and Lavocat (1974) involving a South Atlantic crossing. This is because of the seemingly Asian origin of the group and the late divergence of African and South American families. If they had migrated via the Laurasian continents there should have been greater morphological variability than is apparent, because of the organisms being obliged to traverse a great variety of environments and latitudes. This interpretation has been supported by an important discovery (Wyss *et al.* 1993) of a caviomorph rodent in strata in Chile that could be as old as Upper Eocene, thereby putting back the arrival of these rodents in South America by 10 million years. The morphological affinities of the fossil are with African, not North American relatives.

Platyrrhine monkeys have been found in the Upper Oligocene of Bolivia (Aiello 1993). If primates arrived in South America in Palaeogene times they would have had to cross a major water barrier. In the past, parallel evolution has been invoked to account for the similarities of the platyrrhines and catarrhines but this is thought unlikely by Aiello because modern research on both comparative morphology and biochemistry indicates a common ancestry. Aiello agrees with George that the most likely biogeographic hypothesis is of a direct link between South America and Africa across the South Atlantic. The anthropoid primates probably originated somewhere in the southern continents and platyrrhines in Africa. The early existence of euprimates in Africa facilitated the dispersal of ancestral lemurs to Madagascar. This is consistent with the diversity of the Fayum anthropoid fauna, with the Fayum parapithecids being the best candidates for the immediate ancestors of the platyrrhines. This hypothesis has been strengthened by the newly discovered pre-Oligocene African primate fauna.

The evidence strongly suggests the existence of a Palaeogene transoceanic sweepstakes route between Africa and South America, and presumably also a similar route between Africa and Madagascar. As mentioned earlier, islands strung along the Rio Grande–Walvis Ridge seem the likeliest candidate, because it is hard to conceive how even small mammals could undertake such a long marine journey without some island hopping.

SUNDERING OF THE PATAGONIA–ANTARCTICA–AUSTRALIA LAND ROUTE

To use Croizat's terminology, there is strong evidence from living biota of a generalized track between South America and Australia via Antarctica. Thus the diverse marsupial fauna of Australia and less diverse one of South America are likely to be significant because this group of mammals is absent from Holarctic faunas, apart from didelphids (opossums) that have migrated to North America from South America in the late Neogene (Tedford 1974). Other vertebrate groups supporting such a generalized track include ratite and galliform birds, chelid turtles, and ceratodontid and osteoglossid fish (Cracraft 1973, 1980). Among the invertebrates is the well analysed case of chironomid midges (Brundin 1988). The animal record is supported by that of the plants. The best known and most cited case is that of the southern beech *Nothofagus*, to which can be added the Proteaceae, a family confined to Australia, southern Africa, and South America. Although plants can disperse more readily than animals there is good evidence that *Nothofagus* cannot disperse effectively across marine barriers (Raven and Axelrod 1974; Raven 1979). With insect and many other animal and plant groups, the relationship between Australia and South America is stronger than either is with Africa (Keast 1973).

The question must next be asked – what does the fossil record show? Marsupials were well established in North America in the Late Cretaceous but the only ones to survive the mass extinction at the end of the period were the didelphids which, by the Late Palaeocene, had extended their range to Europe. Marsupials did not reach Asia, because of the barrier imposed by the Turgai Strait. Late Cretaceous South American marsupials resemble their contemporaries in North America, especially the didelphid *Alphadon*. Tedford (1974) speculates that dispersal from a South American centre could have provided the nucleus for both North American and Australian radiations. The Late Cretaceous and Early Tertiary record of South America suggests a long marsupial history there. There is no record of Late Mesozoic or Early Cenozoic marsupials in Australia. According to Tedford the radiation was probably initiated by invading insectivorous didelphines. Both comparative morphology and serology indicate that all living Australian families are more closely related to each other than to any New World family, suggesting derivation from a common stock. Tedford considers that immigrants dispersed in either the Late Cretaceous and Palaeocene and subsequently became isolated. The first marsupial discovered in Antarctica is a polydolopid, which occurs in the Upper Eocene of Seymour Island, at the tip of the Antarctic Peninsula (Woodburne and Zinsmeister 1984).

With regard to plants, there is abundant *Nothofagus* pollen in the Maastrichtian of Australia, New Zealand, and Patagonia, but no reliable records of Cretaceous or Palaeogene Fagaceae in India or Africa south of the Mediterranean region (Raven and Axelrod 1974). More recently it has become established that there is also a Late Cretaceous–Palaeogene record of *Nothofagus* and austral gymnosperms in Antarctica (Raven 1979) and Proteaceae in Australia (Kemp 1978).

Woodburne and Zinsmeister (1984) have attempted to reconstruct the palaeogeography of the severance of the southern land connection. A continuous link between South America and Australia must have existed in the Late Mesozoic. A rift zone between Australia and Antarctica began to open in the Late Jurassic. Re-evaluation of magnetic anomaly data indicates that deep-sea conditions may have developed as far east as 140° (western Victoria) by about 80 Ma rather than 56 Ma as thought earlier. No marine connection from the Indian to the Pacific Ocean was established, however, until the Eocene. If such a connection was established earlier, it is difficult to account for the dissimilarity of Late Cretaceous–Palaeocene shallow-water molluscan faunas between south-western Australia and south-eastern Australia–New Zealand. Geological and palaeontological data suggest a

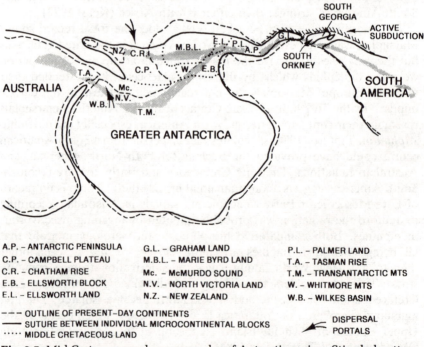

A.P. – ANTARCTIC PENINSULA	G.L. – GRAHAM LAND	P.L. – PALMER LAND
C.P. – CAMPBELL PLATEAU	M.B.L. – MARIE BYRD LAND	T.A. – TASMAN RISE
C.R. – CHATHAM RISE	Mc. – McMURDO SOUND	T.M. – TRANSANTARCTIC MTS
E.B. – ELLSWORTH BLOCK	N.V. – NORTH VICTORIA LAND	W. – WHITMORE MTS
E.L. – ELLSWORTH LAND	N.Z. – NEW ZEALAND	W.B. – WILKES BASIN

– – – OUTLINE OF PRESENT–DAY CONTINENTS
——— SUTURE BETWEEN INDIVIDUAL MICROCONTINENTAL BLOCKS
····· MIDDLE CRETACEOUS LAND

DISPERSAL PORTALS

Fig. 9.5 Mid Cretaceous palaeogeography of Antarctic region. Stippled pattern indicates possible overland dispersal routes. Simplified from Woodburne and Zinsmeister (1984).

land bridge along the South Tasman Rise between North Victoria Land and south-eastern Australia (Fig. 9.5). The final separation, with the onset of deep-water circulation between the two oceans, took place at about 38 Ma. Thus until this time the dispersal of land organisms was possible.

Geological and palaeontological evidence indicates that by the latest Cretaceous New Zealand was completely separated from Antarctica. By the Late Palaeocene it had migrated a considerable distance to the north, parallel to Australia. The absence of any fossil or endemic extant mammals in New Zealand may indicate that Greater Antarctica was not populated by mammals prior to the latest Cretaceous separation of New Zealand. The existence of mammals on the Antarctic Peninsula shows that it was closely linked to South America in the Late Cretaceous and Palaeogene. However, the presence of marine sediments in the southernmost Andes indicates that a shallow trans-Andean seaway connected the Pacific with the South Atlantic at this time, and may have persisted until the Neogene.

The Antarctic Peninsula is believed to comprise a number of discrete or semi-discrete microcontinental blocks (Dalziel and Elliott 1982). The Scotia Arc had not started to open and the northern tip of the peninsula was adjacent to the southern tip of South America. The South Georgia and South Orkney blocks became detached during the Palaeogene and gradually migrated eastwards. Thus faunal exchange became increasingly unlikely from about Oligocene times onwards. Although large areas of Marie Byrd and Ellsworth lands are currently below sea level, it is likely that most of this part of Lesser Antarctica was above sea level in the Late Cretaceous. The close proximity of New Zealand and the similar ages of tectonic episodes suggest that they formed one orogenic belt, while the presence of dinosaurs in New Zealand is an indication that some land connection with Gondwana may have existed as late as the Late Cretaceous.

Barriers to overland dispersal developed in the following order: 1) Lesser Antarctica–New Zealand; 2) Greater Antarctica–Australia; 3) Antarctic Peninsula–Greater Antarctica; 4) southern South America–Antarctic Peninsula. By the Late Eocene or Early Oligocene dispersal would have been very improbable, especially to and from Australia. The sharply different marsupials of Australia and South America, and the near certainty of a Palaeocene, if not Late Cretaceous, radiation of South American didelphids and borhyaenids, suggests that faunal continuity was very limited, with only didelphids and dasyurids shared. The absence from New Zealand may mean that marsupial dispersals between Greater Antarctica and Australia were limited to the latest Cretaceous.

Climate

Fossil plants and associated faunas indicate mild, essentially frostless climates in polar regions during the Palaeocene and Eocene with mixed conifer-hardwood and deciduous-hardwood forests in the Arctic and forests in austral regions dominated by broad-leaved evergreen dicotyledons and evergreen conifers, including araucarians and podocarps. Because of smaller land areas towards the South Pole dry belts were very restricted to absent. Hence the deciduous habit was not commonly developed as evergreens shifted poleward (Axelrod 1984).

Before the plant record is discussed more fully mention should be made of the important record of Early Eocene vertebrates discovered quite recently on Ellesmere Island, in the Canadian Arctic Archipelago. The vertebrates include the tortoise *Geochelone*, varanid lizards, and the alligator *Allognathosuchus* (Estes and Hutchinson 1980). Aquatic varanids occur today only in tropical habitats. Alligators can endure slightly lower temperatures than varanids and *Geochelone* as they can escape the effects of mild freezes by the buffering nature of aquatic habitats. A warm climate, without winter freezing, is confirmed by the mammal evidence, which is consistent with a forested environment marginal to and including coal swamps (McKenna 1980).

Plants have for a long time been regarded as the best means of determining Tertiary temperatures. In North America highly informative results have been obtained from analysis of fossil leaves (Wolfe 1978). An excellent correlation is claimed by Wolfe to exist between the type of leaf margin and climate, with the percentage of species with entire-margined leaves, i.e. lacking lobes and teeth, increasing systematically with temperature. In areas of high mean annual temperature and precipitation the leaves tend to be entire-margined, evergreen, large and with a leather-like (coriaceous) texture, a high proportion of 'drip tips', and a tendency towards palmate venation. The mean annual range of temperature is more difficult to infer than the mean annual temperature but can be estimated in some cases, for instance by the proportion of microphyllous to notophyllous leaves. Some scepticism about this use of leaf margin physiognomy has been expressed and it should probably be used for indicating general climatic conditions rather than as a precise climatic indicator (Boyd 1990).

Wolfe (1978, 1980) applies these various criteria to the interpretation of climatic change through the Tertiary in the northern hemisphere, the most comprehensive data coming from the north western United States. In Palaeocene and Eocene times tropical rain forest, and the 25° isotherm, extended 20–30° poleward of the present northern limit. There was a major drop from the Eocene to the Oligocene in mean annual temperature of about 12–13° in Alaska and Washington State. The mean annual range of temperature increased from 3–5°C in the Mid Eocene to 21–25°C in the Oligocene. Arctic Palaeocene floras are better known than those of the Eocene. South Alaskan forests were dominated by broad-leaved evergreens, including palms and cycads, while West Greenland had broad-leaved evergreen forests together with broad-leaved deciduous trees and minor proportions of conifers. A Late Palaeocene–Early Eocene flora in North Greenland indicates a mixed evergreen and deciduous coniferous and broad-leaved forest representing either warm temperature conditions or moderate temperatures with rare frost (Boyd 1990). Boyd considers that leaf physiognomy gives much cooler conditions than comparisons with living relatives. Leaf physiognomy can be affected by low light levels altering leaf margins, leaf size, and thickness, as evident in the tropical forest understory. This may explain the discrepancy.

Palaeobotanical and palynological results from western Europe confirm those from North America in indicating climatic cooling from the Eocene to the Oligocene (Chateauneuf 1980; Collinson *et al.* 1981; Hubbard and Boulter 1983). The Mid Eocene flora was predominantly tropical, with the diversity of tropical taxa reaching a maximum in the late Lutetian. By the Late Eocene many tropical taxa had become extinct, with the dense forests being reduced and replaced by taxodiaceous swamps and reed marshes, with patches of woodland or forest.

The best southern hemisphere data come from Australasia, mainly from palynology (Kemp 1978). Kemp confirms the basic conclusions of earlier workers that Early Tertiary climates of Australia were characterized by greater humidity and were warmer than today. Rainforest vegetation dominated the landscape across the southern margin of the continent, locally spreading inland to central Australia; these inland forests could have been extensive or restricted. There was a major fall in temperature and associated decrease in floristic diversity at the end of the Eocene, but extensive rainforest persisted in the Oligocene. The high fern diversity, the types of tree and the presence of epiphytic fungi point overwhelmingly to high humidity. Coastal vegetation existed in Antarctica, so that any ice must have been confined to alpine-type glaciers.

Wolfe (1978, 1980) has argued that broad-leaved evergreen trees cannot survive under present light levels poleward of 49° latitude. This has led him to suggest that the obliquity of the Earth's axis was less in the Early Palaeogene than today. This interpretation has provoked a number of criticisms, with several palaeobotanists pointing out that temperature, not light, is the prime limiting factor on plant growth in polar regions (Axelrod 1984; Creber and Chaloner 1985; Boyd, 1990). Boyd observes that evergreens require less photosynthesis than deciduous trees at temperatures lower than the optimal, and could therefore survive the lower incidence of light in high latitutdes. In addition to these empirical objections, Barron (1984) has shown that decrease in the angle of inclination would result in less light being received at the poles than now. Modelling of the resulting climate produced substantially lower temperatures than are able to support the flora and fauna that lived in high latitudes in the Early Palaeogene.

The pronounced decline of global temperatures in the Late Eocene, accelerating at the end of the epoch, is thought to have been the prime factor in the increase in extinction rate of mammals (Prothero 1989). The reduction in vertebrate diversity in the Palaeogene need not exclusively be due to this factor as indicated by the study of aquatic vertebrates in the western United States by Hutchinson (1982). The highest diversities of turtles, crocodiles, and champsosaurs are in the Early to Mid Eocene and Mid Miocene, with a notable fall in the Late Oligocene. This fall is unlikely to be the result of lowered temperatures, because of the high diversity and abundance of contemporary large terrestrial turtles, which would have been vulnerable to prolonged winter temperatures below about 13°C. Indeed, lowered Pleistocene temperatures are probably the reason for the decline of large tortoises. Instead, Hutchinson argues that the aquatic vertebrates were restricted or eliminated by a major increase in aridity between 20 and 30 Ma, leading to a reduction in the amount of surface waters.

DISTRIBUTION OF MARINE ORGANISMS IN RELATION TO PALAEOGEOGRAPHY AND CLIMATE

As a result of the deep ocean drilling programme that began a quarter of a century ago there is a rich record of Palaeogene planktonic and benthic microorganisms, notably foraminifers, ostracods, and coccoliths, in stratigraphic sequences which are generally more complete and precisely datable than those of the neritic zone. In conjunction with oxygen isotope data they provide the best record of water temperatures. Biogeographically the most interesting features concern the Tethyan zone and the Southern Ocean adjacent to Australia, Antarctica, and Patagonia.

The Tethyan zone

The ancient Tethyan seaway separating the northern and southern continents continued into Palaeogene times, with free marine continuity along its length not being interrupted until the beginning of the Neogene. In consequence there is little endemism throughout the Old World. Some of the most characteristic faunas were the larger benthic foraminifers, the most familiar of which are the nummulitids, which can occur in rock-forming abundance as in Egypt (Adams 1973; Hottinger 1973). The larger foraminifers were not confined to the Tethyan zone and were able to migrate quickly over vast areas. Rafting is considered by Chapronière (1980) to have been a likely mode of transport, because there is uncertainty about whether or not they had a planktonic larval phase. If they did it is puzzling why they are not recorded in deep-sea sediments. Dispersal must have been aided, if not controlled, by continuous coastlines or by island chains separated by shallow seas.

Other important elements include hermatypic corals, which did not re-establish proper reefs after the end-Cretaceous extinctions until the Mid Eocene (Winterer 1991), and echinoids (Rosen and Smith 1988; Ghiold and Hoffman 1986). The parsimony endemicity analysis of Rosen and Smith indicates that, although the definitive separation of Mediterranean from Indian Ocean coral and echinoid faunas took place in the Miocene, marine biotas had become differentiated at generic level long before, at least throughout the Palaeogene.

Ghiold and Hoffman undertook a biogeographic study of the clype-asteroid echinoids (sand dollars and sea biscuits) which rapidly expanded from their origins in the Palaeocene to achieve a global distribution in the Eocene. At present there are twelve extant genera and over a hundred species. The biogeographic data were subjected to cluster analysis, using an unweighted pair-group method with the Jaccard Coefficient as the similarity metric. The fossil record solves some, but not all problems inherent in the biogeographic cladogram based on living taxa. Thus the extensive Mediterreanean fossil record unequivocally puts this sea in close relationship with the Indian Ocean and West Pacific. The concept favoured by some vicariance biogeographers of ancestral distributions being all inclusive and having subsequently undergone vicariance only, is rejected. In addition to early, or primary, dispersal there were secondary migrations as well as vicariance events, involving possibly diffusion or even chance dispersal in addition to secular migration.

The Southern Ocean

Planktonic biogeography of the Southern Ocean during the Cenozoic has been studied by Kennett (1978). Nearly all the evolution of calcareous plankton occurred outside the Southern Ocean, with subsequent migration there, such that there is virtually no endemism. Planktonic foraminifers of the Subantarctic Eocene have relatively strong affinities with those of temperate regions, but biogeographic differences exist between various sectors of the Southern Ocean as a consequence of isolation preceding the development of the Circum-Antarctic Current. Palaeocene and Eocene calcareous nannofossils show higher diversity than Oligocene assemblages. The radiolarian and diatom record is relatively poor. Close to the Eocene–Oligocene boundary a dramatic change took place. Since that time Antarctic planktonic foraminifers exhibit a distinct polar aspect, with simple morphotypes and low diversity. Oligocene plankton indeed tend to exhibit low diversity and relatively cosmopolitan assemblages throughout the world.

With regard to the molluscan neritic fauna of the southern circum-Pacific, many endemic taxa have been recognized. Of 66 Early Palaeocene mol-luscan genera in New Zealand 71 per cent are known only from this region (Zinsmeister 1982). The families Lahilliidae and Taiomidae and the genus *Struthioptera* disappeared from Australia and New Zealand after the Palaeocene but survived well into Mid Tertiary times in Antarctica. According to Zinsmeister, palaeoaustral species in Australia and New Zealand might well have declined as a consequence of experiencing warmer conditions and competition from invading Indo-Pacific species as these regions moved northwards. By the Mid Eocene (Bartonian) warm-water Indo-Pacific elements made their first appearance in New Zealand; these include the gastropod families Conidae, Cypraeidae, Mitridae, and Harpidae. There was a decrease in the palaeoaustral component from 70 per cent in the Early Palaeocene to less than 10 per cent in the Mid Eocene.

The Antarctic molluscan record, extending up to the Lower Oligocene, is almost entirely based on Seymour Island. While the faunas had to adapt to cooler conditions they did not have to cope with the more drastic changes imposed on the Australasian faunas. Thus palaeoaustral elements were still an important component (31 per cent) even in the Late Eocene, in contrast to the situation in New Zealand. There was a large diversity drop across the Eocene–Oligocene boundary, from 65 to 20 genera.

The presence of the crab *Lyreidus antarcticus* in the Eocene of Seymour Island suggests a close faunal association with New Zealand, where the genus is also known in deposits of this age. All these crabs occur in very shallow-water facies, whereas living species of the genus are found primarily in fine-grained outer shelf and slope sediments (Feldman and Zinsmeister 1984). The similarities between both molluscs and *Lyreidus* in the Eocene of Antarctica and the Early Palaeogene of New Zealand suggest that these organisms might very well have had their origins in New Zealand and dispersed eastwards. This pattern of dispersal would be consistent with the establishment of the Weddelian Province (Zinsmeister 1982).

Before the Mid Oligocene Australia–New Guinea–New Zealand were part of the Indo-Australian Plate well separated from the Solomon–New Hebrides–Fiji part of the Pacific Plate and Java–Borneo part of the Asian Plate. By the Mid to Late Oligocene the New Guinea and Java–Borneo areas were close enough to permit intermigration of larger foraminifers to occur with *Cycloclypeus*, *Lepidocyclina*, and *Miogypsina* invading the Australia–New Guinea areas (Chapronière, 1980). This is a further biogeographic response to the northward migration of the Australasian region away from Antarctica.

Climate

The most comprehensive palaeobiogeographic analysis bearing on climate is that of Haq *et al.* (1977; see also Boersma *et al.* 1987) on the calcareous plankton of the Atlantic. For Quaternary times, the temperature tolerances of species are known because they are still extant, but for earlier times this is a problem. To resolve this problem Haq *et al.* have attempted to map the geography of fossil assemblages. Once their spatial and temporal distributions are known, assemblages that show relative restriction to either high or low latitudes can be used as palaeoclimatic indicators. Fluctuations in the biogeographic patterns of assemblages through time can then be interpreted in terms of dynamic changes in the climatic–oceanographic system. For example, spreads of low-latitude assemblages are likely to signify warming episodes and spreads of high-latitude assemblages cooling episodes. Nannoplankton are photoautotrophic, living only in surface waters, while planktonic foraminifers are more vertically differentiated, occurring in different depth zones. In the analysis by Haq *et al.* raw census data are used to determine assemblages by Q mode factor analysis.

Fig. 9.6 summarizes the major nannofloral and foraminiferal migration patterns through the Palaeogene. Four pronounced cooling episodes were inferred for the Mid Palaeocene, Mid Eocene, earliest Oligocene, and Mid Oligocene, and a pronounced warming episode in the Late Palaeocene–Early Eocene together with a milder one in the Late Oligocene. The correlation with oxygen isotope data is reasonably good, though the latter do not record the Mid Palaeocene cooling event.

Benthic foraminifers from intermediate water depths (100–1000 m) have been studied by Kaiho (1992). He recognizes two Late Eocene provinces: 1) North Pacific; 2) South Pacific–Tethyan–Atlantic. Biogeographic divergence may have occurred in the Mid Eocene as a result of intermediate waters originating at high latitudes due to global cooling. These Late Eocene faunas did not mix thereafter, presumably because of closing of Tethys and the presence of a deep-sea basin between the North and South Pacific. Kaiho's study is the first on benthic foraminifers involving a comparative study of specimens by a single investigator, and demonstrates lower regional endemicities than have been recognized hitherto.

The most profound cooling episode was across the Eocene–Oligocene boundary. From oxygen isotope data from benthic and planktonic foraminiferans in the Southern Ocean, Shackleton and Kennett (1975) established a fall of 5°C in bottom and 7°C in surface waters. Isotopic data indicate that Australia and New Zealand waters experienced a general climatic warming as the region moved away from the Antarctic, with a downturn only at the Eocene–Oligocene boundary (Zinsmeister 1982).

The term psychrosphere was coined by Brunn (1957) for the cold and denser layer of the modern two-layer ocean, the other layer being the thermosphere. According to Benson (1975) the benthic record of ostracods indicates that the psychrosphere came into existence suddenly at the Eocene–Oligocene boundary, associated with the creation of cold Antarctic bottom waters. Deep-sea faunas from times before the event are

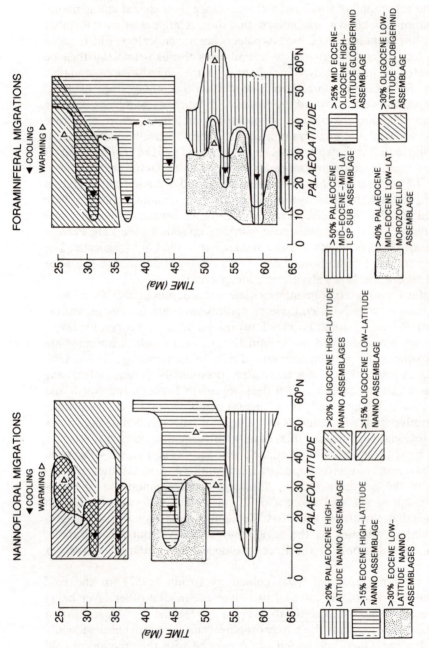

Fig. 9.6 Summary of the major nannofloral and foraminiferal migrationary patterns through the Palaeogene. After Haq et al. (1977).

taxonomically different from those after the event. Water temperature or some other factor changed the basic architectural forms of the ostracods.

As with the terrestrial mammals, there was an increase in extinction rate of planktonic and benthic organisms in the Late Eocene, presumably related to climatic cooling. According to Keller (1983) there were relatively few extinctions among planktonic foraminifers across the Eocene–Oligocene boundary, simply an increase in dominance of cold-water taxa. The most severe extinction event among the calcareous nannoplankton was across Bartonian–Priabonian boundary (Aubry 1983). As noted earlier, the neritic record is much poorer than that from the deep sea, but Hansen's (1987) thorough work on American Gulf Coast molluscs demonstrates a pattern parallel to that recognized for calcareous plankton. Mention should also be made of the classic work of Durham (1950), who demonstrated that Early Palaeogene neritic faunas of the Pacific coast of North America have a subtropical to tropical aspect in much higher latitudes than today.

A major extinction of deep-sea benthic foraminifers has been recognized at the end of the Palaeocene which left planktonic foraminifers little affected. It coincides with a remarkable oxygen and carbon isotope excursion in Antarctic waters, indicative of rapid global warming and concomitant oceanographic changes. This dramatic event, which probably lasted no longer than a few thousand years, caused extinction probably because of the rapid temperature increase and associated reduction of oxygen concentrations (Kennett and Stott 1991).

10
Neogene

The Neogene comprises the Miocene, Pliocene, Pleistocene, and Holocene epochs. Since the Holocene began only about 10 000 years ago it is commonly grouped together with the Pleistocene as the Quaternary. The biogeographic literature on the Quaternary vastly exceeds that on the Miocene and Pliocene, and deals almost exclusively with extant species. Here the Quaternary is treated in a more balanced way as a small, albeit highly interesting, part of the Neogene, and emphasis is placed, as in previous chapters, on what is indicated by the changing distributions of fossil taxa.

MAJOR GEOLOGICAL EVENTS

Sea-floor spreading continued in all the major oceans and the Atlantic and Indian Oceans widened further. Features of particular biogeographic significance included the closure of Tethys in the Middle East, the collision of Australia–New Guinea with Indonesia and the emergence of the Panama Isthmus providing a land link between North and South America. Palaeobiogeographic data are of vital importance in determining the precise timing of such events, and provide the two best examples known of complementarity (see Chapter 2).

The complex Neogene history of the region surrounding the present Mediterranean is reviewed by Rögl and Steiniger (1984). The most dramatic changes took place in the Miocene and involved the formation of what is known as Paratethys, extending from the Rhone Valley and Bavaria to the Caspian and Aral Seas. Collision of Africa with Eurasia took place in the Early Miocene in the Middle East region, permitting cross migration of land mammals but severing the marine link between the Mediterranean and Indian Ocean (Fig. 10.1). The land-locked basins of Paratethys were much altered by the Africa–Eurasia collision, and a restriction of open marine circulation caused Early Miocene precipitation of evaporites in the Rumanian–Ukrainian Carpathian foredeep. A low-salinity sea became established soon afterwards in the eastern Paratethys, in the Mid Burdigalian, and an endemic molluscan fauna became established there.

A Mid Miocene marine transgression spread all over the Mediterranean

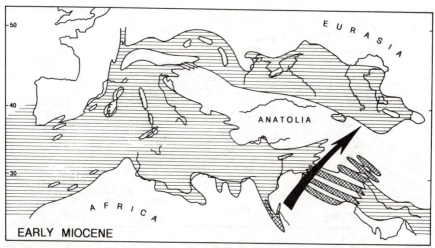

Fig. 10.1 The Eurasian–African collision. The closure of the seaway to the Indo-Pacific in the Middle East permitted the first large mammal migration between Africa and Eurasia in the Early Miocene. Horizontal lines signify normal marine conditions, cross hatch pattern indicates continental regimes within seaway. Modified from Rögl and Steininger (1984).

and Paratethys, reopening a seaway to the Indo-Pacific, but a disruption of marine connections in the Ponto-Caspian region led to a second episode of reduced salinity with endemic molluscs. This episode was only short-lived and the entire Paratethys was again flooded by sea later in the Mid Miocene, but the connection with the Mediterranean had been closed, and differences between the marine biotas developed. In Serravalian times (about 14 Ma) fully marine conditions ended in Paratethys and salinity was reduced to 16–30 per cent at the end of the Sarmatian (late Middle Miocene), as marked by the disappearance of stenohaline organisms. The partly endemic Sarmatian assemblages have a few euryhaline taxa in enormous numbers, including elphidiids and miliolids among the Foraminifera and cardiids, mactrids, cerithiids, and trochids among the Mollusca.

The most striking event took place at the end of the epoch, in the Messinian. As indicated by the presence, as proved by boreholes, of vast quantities of evaporites, it is evident that the Mediterranean became a hypersaline, land-locked sea substantially below ocean level, while Paratethys split up into other land-locked seas of very low salinity (Fig. 10.2). At the end of the Messinian brackish-water sediments of the so-called Lago Mare facies were very widespread in the Mediterranean region. This marked the time of mammal migrations to Mediterranean islands. Marine flooding took place in the Early Pliocene, when the Red Sea became independent and the Aden Strait opened into the Indian Ocean.

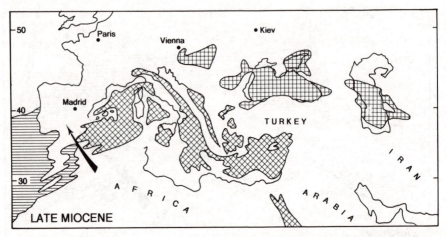

Fig. 10.2 The Messinian (latest Miocene) salinity crisis in the Mediterranean region. Horizontal lines signify normal marine conditions, oblique cross hatch pattern abnormally high salinity, vertical and horizontal cross hatch pattern abnormally low salinity. Modified from Rögl and Steininger (1984).

Collision between Australia–New Guinea and the Sulawesi and adjacent parts of Indonesia is dated as Mid Miocene (about 15 Ma) by Audley-Charles (1981). Seram and Timor on the northern Australian margin collided at 5 and 3 Ma respectively as Australia continued its northward drift (Fig. 10.3). Extensive land could have been exposed between Australia and Sulawesi by the latest Miocene or Early Pliocene, allowing a migration corridor. This could be relevant to Estes' (1983) claim of strong similarities between the living lizard populations of Australia and South-east Asia.

There is a disparity between the vertebrate faunas of Australia and South-east Asia, with the line of division being drawn by what has long been called Wallace's Line (Whitmore 1981*a*). The precise location of this line has been disputed, and Wallace himself changed his mind about where to place Sulawesi. Some have preferred Weber's Line, based on mammals and molluscs, which conforms more or less to the 100 fathom contour between Sulawesi and the Moluccas (George 1981). Unlike vertebrates and molluscs, Wallace's Line is unimportant for the great majority of plants (George 1981; Whitmore 1981*b*). Wegener (1924) considered the sharp contrast between Wallace's Oriental and Australian Regions *prima facie* evidence in support of continental drift, with two hitherto geographically isolated regions coming into juxtaposition in the recent geological past.

With regard to the Central American and Caribbean region, no significant change in the gross geological relations took place after the Oligocene (Fig. 9.3). A critically important emergence event in the Panama area had the effect of allowing free communication of land organisms between the Americas, while severing even shallow marine links between the Pacific and Caribbean. The biogeographic consequences of this will be considered more fully later. A recent stratigraphic analysis of the Neogene of the Atrato Basin in north-western South America has allowed a detailed reconstruction of the evolution of the Panama seaway (Duque-Caro 1990).

The Atrato Basin, like other Neogene coastal basins of south Central and northern South America, has three major stratigraphic sequences:

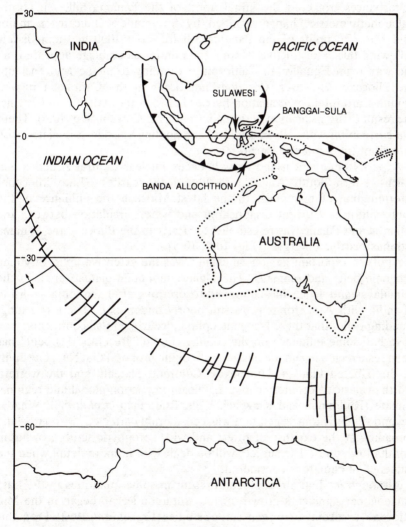

Fig. 10.3 Reconstruction of Australia and South-east Asia for the Late Miocene. Adapted from Audley-Charles *et al.* (1988).

(1) Upper Oligocene–Middle Miocene; pelagic and hemipelagic sediments, signifying open and deep oceanic environments;

(2) Upper Middle Miocene–Lower Pliocene; hemipelagic and terrigenous sediments, of mid bathyal to neritic environments;

(3) Post–Lower Pliocene; mostly fluvial and lacustrine sediments.

Prior to the Mid Miocene there were well aerated, deep oceanic conditions, with free active water circulation. Early in the Mid Miocene tectonic disturbances triggered the initial uplift of the Panama Sill. An abrupt palaeobathymetric change signified by foraminifers indicates a major Mid Miocene uplift of the Panama Sill to about 1000 m. Immediately following this, a distinctive benthic foraminifer assemblage appeared, all the way from Equador to California, extending from the Miocene into the Pliocene. However, this does not occur in the Caribbean region. Duque-Caro infers a circulation barrier between the Atlantic and Pacific, the result of intensification of the cool California Current (Fig. 10.4). There is a correlation with falls in sea level and cooling episodes associated with growth of the Antarctic icecap.

A partly emergent island mass between nuclear Central America and north-western South America facilitated the earliest (Late Miocene) intermingling of mammals. In the latest Miocene the influence of the cool California Current disappeared and water circulation between the Atlantic and Pacific was re-established. Early in the Pliocene the Panama Isthmus became emergent (Figs 10.4, 10.5).

Another important geological change was the extensive uplift of young orogenic belts and plateaus. The highest mountain-plateau area, in the Himalayas and Tibet, has continued to rise long after the initial collision of India with Asia, for reasons still poorly understood. There is a long-standing belief that these Neogene uplifts, occurring in all continents, may have had some influence on the cooling climate. Recently this belief has been resurrected in climate modelling with the proposal that Neogene uplift of the Tibetan Plateau, Himalayas, Colorado Plateau, and the western North America Cordillera had a significant impact on global and regional climate (Ruddiman and Kutzbach 1989; Ruddiman *et al.* 1989). What is beyond doubt is that vegetation was considerably affected, with grassland spreading at the expense of forest and with xerophytic scrub growing in rainshadow zones. The subject will be dealt with in more detail when the influence of climate is considered.

Although ice had already been present in some quantities in the Late Palaeogene, significant growth of the Antarctic icecap began in the Mid Miocene, according to oxygen isotope studies (Frakes *et al.* 1992). From this time onwards the world was subjected to considerable climatic fluctuations, especially in the Quaternary, as polar icecaps expanded and retreated. One

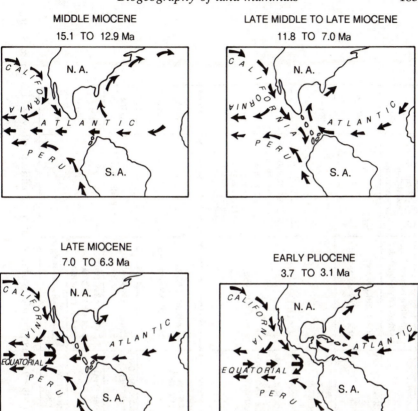

MIDDLE MIOCENE
15.1 TO 12.9 Ma

LATE MIDDLE TO LATE MIOCENE
11.8 TO 7.0 Ma

LATE MIOCENE
7.0 TO 6.3 Ma

EARLY PLIOCENE
3.7 TO 3.1 Ma

Fig. 10.4 Neogene palaeoceanographic and palaeogeographic evolution of the central American region. Hypothetical surface circulation resulting from the disruption of the warm Caribbean flow and intensification of the cool California current. Simplified from Duque Caro (1990).

major consequence of this was a series of glacioeustatic sea-level changes (Hallam 1992). At times of low sea level in epicontinental seas, such as the north-west European Shelf, the Bering Strait, and Sunda Shelf (of Indonesia and Malaysia), land links became established, facilitating intermigrations of terrestrial organisms.

BIOGEOGRAPHY OF LAND MAMMALS

The mammals are the only terrestrial organisms for which there is a rich biogeographic record from which changing connections and migrations between continents can be inferred. During the Neogene all continental

AGES	TECTONICS	PALAEOCEANOGRAPHY	PALAEOBIOGEOGRAPHY		HIATUSES	CLIMATE
			MARINE	TERRESTRIAL		
EARLY PLIOCENE 3.7–3.1Ma	UPLIFT AND COMPLETE EMERGENCE OF THE PANAMA ISTHMUS.	SHALLOW WATER CONNECTION CLOSED. SEA LEVEL DROP.	DIMINISHING OF CARIBBEAN AFFINITIES	ONSET OF THE GREAT AMERICAN INTERCHANGE	NH 8	COLD
LATE MIOCENE TO EARLY PLIOCENE 6.3–3.7 Ma	SHALLOWING TO WATER DEPTHS LESS THAN 50m	SHALLOW WATER CONNECTION RESTRICTED. ANOXIC BOTTOM. SEA LEVEL RISE.	CALIFORNIAN AND CARIBBEAN AFFINITIES			WARM
LATE MIOCENE 7.0–6.3 Ma	UPLIFT TO WATER DEPTHS LESS THAN 150m	SHALLOW WATER CONNECTION RESTRICTED. END OF THE CALIFORNIA CURRENT ACTION. SEA LEVEL DROP.			NH 6	COLD
LATE MIOCENE 8.6–7.0 Ma	SHALLOWING TO UPPER BATHYAL DEPTHS.	SHALLOW WATER CONNECTION OPEN. COOL AND WELL AERATED SURFACE WATERS OF THE CALIFORNIA CURRENT. ANOXIC BOTTOM. SEA LEVEL RISE.	CALIFORNIAN AFFINITIES	EARLIEST RECORD OF INTERMINGLING 9.3 TO 8.0 Ma PROCYONIDS (RACCOONS) AND MEGALONYCHIDS (GROUND SLOTHS)		COOL
LATE MIDDLE TO LATE MIOCENE 11.8–8.6 Ma	APPARENT STABILITY					
MIDDLE MIOCENE 12.9–11.8 Ma	UPLIFT TO UPPER MIDDLE BATHYAL DEPTHS. INNER BORDERLAND BASINS FORMED. PARTIAL EMERGENCE OF THE PANAMA ISTHMUS.	CLOSING OF INTERMEDIATE WATER CONNECTION AND ONSET OF THE CALIFORNIA CURRENT ACTION. SEA LEVEL DROP.	ABRUPT END OF CARIBBEAN AFFINITIES	NO INTERCHANGE	NH 3	COLD
MIDDLE MIOCENE 15.1–12.9 Ma	APPARENT STABILITY	INTERMEDIATE AND SHALLOW WATER CONNECTION OPEN. SEA LEVEL RISE.	CARIBBEAN AFFINITES			WARM
EARLY–MIDDLE MIOCENE 16.1–15.1 Ma	UPLIFT TO LOWER BATHYAL DEPTHS	CLOSING OF THE DEEP WATER CONNECTION. SEA LEVEL DROP.	CARIBBEAN AFFINITIES. RADIOLARIANS DISAPPEAR IN THE CARIBBEAN.		NH 2	COLD
EARLY MIOCENE 23.7–16.2 Ma	APPARENT STABILITY	DEEP WATER CONNECTION OPEN. SEA LEVEL RISE.	CARIBBEAN AFFINITIES			WARM

Fig. 10.5 Neogene, tectonic, palaeoceanographic, and palaeobiogeographic evolution of the Pacific north-west corner of South America. After Duque-Caro (1990).

masses had their own distinctive mammals and much of the taxonomic diversification throughout mammalian history has been the result of convergent evolution of similar ecomorphs from different taxonomic sources on different continents (Kurtén 1973). The present diversity is much lower than in the Mid Tertiary, with the endemic faunas of Africa and South America having suffered major diversity reductions since Holarctic immigrations (Janis and Damuth 1990). The most important intercontinental migrations were between North and South America, North America and Eurasia, and Africa and Eurasia.

North and South America

What has become known as the Great American Biotic Interchange is comprehensively dealt with in Stehli and Webb (1985), Marshall (1988), and Webb (1991). Two synchronous and reciprocal Pliocene dispersal events are recognized, based primarily on the first known record of each taxon on the continent they dispersed to and on the sequence of glacial maxima, implying low sea level and hence facilitated exchange. The first was at about 2.8–2.6 Ma (Blancan) and the second 2.0–1.9 Ma (Irvingtonian).

There was also an Early Pleistocene dispersal event dated at about 1.4 Ma and some other later events which are not well dated (Marshall 1985). Prior to these times, in the Late Miocene, two genera of ground sloths reached North America and one raccoon genus spread to South America. That no others followed until the main interchange implies the probable presence of water gaps. The acme of the interchange was in the Blancan, and the remarkable conclusion that must be drawn is that as much as half the modern South American mammal genera gained a foothold and evolved there in less than three million years (Webb 1985).

In his classic study, Simpson (1950) first drew attention to the north–south asymmetry, with greater 'success' for northern families. The study of Marshall *et al.* (1982) attempted to demonstrate that the Pliocene interchanges corresponded to predictions from equilibrium theory, but during the Pleistocene interchange became distinctly asymmetric, with far more northern genera populating South America than predicted from theory.

The ecogeography of the interchange has recently been analysed by Webb (1991). Pliocene arrivals in North America of South American taxa consist of ten genera including capybaras, porcupines, armadillos, glyptodonts, and ground sloths. North American immigrants to South America by the end of the Pliocene number ten genera including cricetid rodents, mustelid carnivores, and some artiodactyls. The initial phase of the interchange produced balanced reciprocal movements. By the Early Pleistocene five more South American and eleven more North American genera had crossed. The new South American immigrants in North America included anteaters and opossums. Eleven families did not extend their range to the north and nine to the south. Alternative explanations for this are lack of proximity to the isthmus or temperate habitats.

An important question to consider is 'Did extinctions favour the natives or immigrants?' Five families that entered North America, including ground sloths and glyptodonts, became extinct, as did two that went south, including horses. One North American family that migrated to South America, the Camelidae, became extinct in the homeland. There was a greater extinction of immigrant taxa in the north than in the south, which is consistent with the existence of a widespread North American glaciation. Many South American taxa such as capybaras, tapirs, and vampire bats became extinct in temperate latitudes but survived in the tropics.

The next question is 'Did immigrant taxa replace the natives?' Simpson (1950) originally argued for this but later (1980) backtracked to some extent. According to Webb no-one has yet made a convincing case for competition in North America, but Patterson and Pascual (1972) suggested that the borhyaenids, the largest indigenous South American carnivores, were displaced by placental carnivores from North America. Detailed evidence of complementary distributions in time and space of ecologically related groups is needed to provide a convincing demonstration of biotic interaction. There is some evidence of this for ungulates. During the interchange ungulates of North American origin diversified to a maximum of twenty genera, while those of South American origin dwindled from thirteen to three genera (Fig. 10.6).

Webb puts forward a two-phase ecogeographic model. In the interglacial phase, forests with closed canopies were dominant throughout the American tropics excluding the Andes, while in the glacial phase the forests were mainly replaced by savanna, thereby facilitating free communication for many groups. Many interchange genera are best interpreted as having this preference. Two tests of the model are proposed, based on latitude and phylogeny. The mean southernmost latitude for all North American fossil and Recent genera is 40°S, whereas the mean northernmost latitude for South American genera is 28°N, which is held to support the hypothesis. As regards phylogeny, most differentiation took place just before the interchange. The great American biotic interchange was not as one-sided as some palaeontologists have claimed, because the 'Central American Paradox' is that within the humid tropics the South American biota evidently conquered the Central American part of North America. Glacial conditions in North America evidently favoured massive immigration to South America.

North America and Eurasia

A number of Neogene mammalian dispersal events between the Palaearctic and Nearctic regions have been recognized (Simpson 1947; Repenning

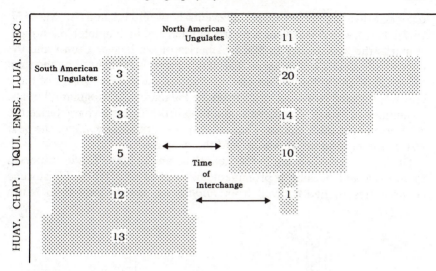

Fig. 10.6 Complementary diversity changes of ungulate genera in North and South America in the Late Neogene. Huay = Huayaquerian, Chap = Chapadmalan, Uqui = Uquian, Ense = Ensenadan, Luja = Lujanian. After Webb (1991)

1967). Faunal migration and exchange evidently took place across the Bering Strait. A 100 m fall of sea level would expose almost the entire area of the Bering–Chukchi continental platform, while a 46 m fall would lead to exposure of a narrow land connection between the Chukotka Peninsula and Alaska (Hopkins 1967). Such low sea-level stands would have occurred on a number of occasions, most markedly in the Pleistocene, as a result of glacioeustasy.

At times of high sea-level stand the link between the Palaearctic and Nearctic became severed, and free communication took place between Pacific and Atlantic marine faunas, an example of complementarity, as will be considered more fully later in this chapter. One such event was in the Late Miocene, which marks a time of isolation not just of mammals but also floras. According to Wolfe and Leopold (1967) the close resemblance between species in warm-temperate broad-leaved deciduous vegetation (Mixed Mesophytic Forest) is best explained by floristic continuity between North America and eastern Asia as late as Mid Miocene times. The complete lack of conspecificity in both Recent and fossil floras seems to indicate effective partitioning in the Late Miocene, related to opening of the Bering Strait. Restoration of a land corridor in the late Neogene would not have restored the status quo because of climatic cooling.

According to Lindsay *et al.* (1984) dispersal between North America and Eurasia was more frequent in the Pliocene than the Pleistocene. They

Neogene

recognize five significant dispersal events between 1.5 and 7.0 Ma (Fig. 10.7). The 4.9–5.2 Ma event is poorly defined and incompletely identified. It marks the appearance in North America of the beaver *Castor* and the microtine rodent *Promimomys*. This latter genus is thought to be the stem for much of an explosive Pliocene radiation of microtines. Waves of these cricetids invaded North America from the Palaearctic at irregular intervals. A number of North American immigrants from Eurasia characterize the 3.7 Ma event in the Blancan. These include the grison *Trigonictis*, the bear *Ursus*, the mastodont *Mammut*, and the deer *Bretzia*.

During the 2.5 Ma event in the Late Blancan the horse *Equus* dispersed to Eurasia and became a prominent member of Old World Pleistocene communities. It had evolved from *Pliohippus*, known only from North

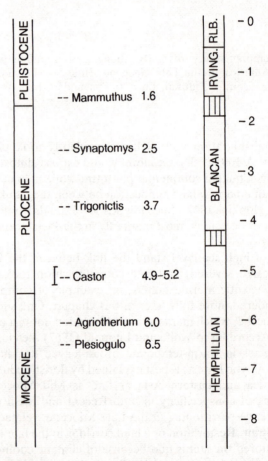

Fig. 10.7 Nearctic–Palearctic mammal dispersal events. Numbers are ages in Ma. Simplified from Lindsay *et al.* (1984).

America (Lindsay *et al.*, 1980). Immigrants to North America at this time included the lemming *Synaptomys* and the bear *Tremarctos*. The 1.6 Ma event, in the Irvingtonian, is marked by the appearance of the elephant *Mammuthus* in North America, following a record in Eurasia of about one million years. Few other mammals invaded from Eurasia at this time.

An increased amount of immigration to North America from Eurasia through time is recognized by Webb and Barnosky (1989). Thus in the Late Miocene the numbers of North American and Eurasian immigrant genera are respectively nineteen and seven, in the Pliocene fourteen and four, in the Early Pleistocene eleven and two and in the Late Pleistocene twenty-one and zero. This is thought to reflect the extent of appropriate biomes in regions adjacent to Beringia. Northern North America, but not Siberia, was extensively covered by ice during the Pleistocene. However, as regards the Pliocene, Flynn *et al.* (1991) maintain that, as more North American taxa are found in Asia, such as in the Yushe Basin of North China, immigration was more balanced in both directions than claimed by Webb and Barnosky.

Although there is no relevant fossil record, our own species was present in Beringia at least 25 ka ago. Cultural, linguistic, and genetic differences between Inuit and Aleuts, and American Indians, indicate clearly that Beringia has seen two waves of human migration. There was probably a time shortly after 25 ka when land connections were simultaneously possible between Siberia and Alaska and Alaska and central North America. An ice-free corridor would then have closed for a 1000 years or more during the Wisconsin glaciation and would not have opened until after the Bering land bridge was severed by rising sea level. This corridor would have allowed a limited period of time for migration southwards of American Indians. The earliest Inuit and Aleuts could have arrived by boat in the early Holocene (Hopkins 1967).

Africa–Eurasia

The Neogene record of Africa also indicates a number of intermittent connections with Eurasia. The first important intermigration took place in the Early Miocene, the time of formation of a land connection in the Middle East (Fig. 10.1). According to Coryndon and Savage (1973), of 70 fossil genera found in the continent 52 are exclusively African. The 74 per cent, endemism is strong but less than in the Oligocene. Out of eleven orders only the Hyracoidea and Tubulidenta are exclusively African. All the 23 immigrant families, including insectivores, creodonts, rhinoceroses, anthracotheres, pigs, and bovids had an earlier history in Eurasia. Three previously endemic African proboscidean families emigrated to Eurasia.

The relationships of African, Arabian, and Indian bovids during the Miocene have been studied by Thomas (1984). Several phases of exchange

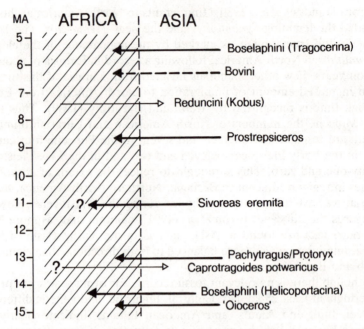

Fig. 10.8 Asian and African bovid immigrants during the Mid and Late Miocene. The arrows show the direction of migration. Adapted from Thomas (1984).

are recognized, with immigration into Africa being more important than the reverse (Fig. 10.8). The micromammal record unequivocally indicates a brief period of faunal exchange between Africa and western Europe at the end of the Miocene, corresponding with the time of the Messinian salinity crisis, when free land communication was available (Thomas *et al.* 1982 and Fig. 10.2; see also Azzaroli and Guazzone 1979; Jaeger *et al.* 1987).

In Pliocene times the differences between North and East Africa became greater, with the Sahara playing an increasingly important role as a faunal barrier. Approximately 50 per cent of the African genera are endemic (Coryndon and Savage 1973). From the Late Pliocene onwards the African fauna became progressively more isolated into separate groups in various parts of the continent. The elephants *Elephas* and *Mammuthus* evolved in Africa from *Primelephas* at about 4 Ma and *Elephas* dispersed to southern Asia at about 2.9 Ma, slightly before the appearance there of *Equus*. The appearance of *Mammuthus* in Europe is approximately coincident with the European appearance of *Equus*, which reached Africa at 1.9 Ma. Apparently *Equus* and *Mammuthus* evolved at approximately the same time in North America and Africa respectively and also dispersed to Eurasia at about the same time (Lindsay *et al.* 1980).

Coryndon and Savage (1973) maintain that Africa was completely isolated from Europe at the end of the Pleistocene. The last mammals of tropical African faunas to enter Europe may have been the Hippopotamidae, Elephantidae, and some large carnivores which managed to survive even in the British Isles until the last glaciation.

The so-called end-Villafranchian dispersal event, about 1 Ma, is interpreted by Azzaroli (1983) as a highly significant event over the whole of Eurasia. This involved the extinction of an 'archaic' Pliocene fauna and the emergence of a new assemblage of modern aspect, including such heavy bovids as *Bos* and *Bison* and the giant deer *Megaceros*. The dramatic change is tentatively related by Azzaroli to significant climatic and vegetational changes prompting adaptation and migration of mammals over large distances, probably triggered by tectonic movements in South and Central Asia.

The record of Neogene mammals from the Indian subcontinent is based mainly on fossils from the Miocene–Pleistocene Siwalik Group and the Lower Miocene Bugti Beds of Baluchistan. The Siwalik Group is most significant in yielding the hominoids *Ramapithecus, Sivapithecus*, and *Gigantopithecus*, while the Bugti Beds have the earliest South Asian record of proboscideans, indicating immigration from Africa. While the Early and Mid Miocene rodent faunas of Europe include many cricetids the Bugti Beds lack this group, confirming that South Asia was probably not the centre of origin for cricetids appearing suddenly in the Middle Tertiary deposits of Europe. Cricetids appear between deposition of the Bugti and succeeding Murree Beds (Jacobs *et al.* 1981).

The rich Siwalik fauna is more or less distinct and contains endemic elements, and is regarded by Barry *et al.* (1991) as an isolated precursor of the modern Oriental Province. Isolation was most complete in the earliest Miocene but subsequently exchange with other regions became increasingly frequent and four phases of greater and lesser isolation are recognized by Barry and his co-workers. The first appearance of bovids and tragulids, together with various rodents, indicates exchange with Africa and Europe. It appears to have been brief and was concluded by 18 Ma. The turnover at 7.5 Ma altered the taxonomic composition of the fauna considerably, making it more similar to contemporary faunas in northern and western Europe. This transformation implies a Late Miocene restriction of the Oriental Province.

Abnormal island faunas

The tendency of small herbivorous mammals, such as rodents, to enlarge, and carnivores and ungulates to dwarf on islands seems to have fewer exceptions than any other ecotypic rule in animals (Roth 1992). 250 gram (the weight of a small rabbit) is the fulcrum, and the rule has been confirmed for a number of Pleistocene and Recent island faunas. Gigantism in small mammals is probably a consequence of immigrant selection and release from competition, while dwarfism in large mammals probably results from resource limitation (Lomolino 1985).

Some excellent examples are provided by Pleistocene fossils collected from Mediterranean islands (Sondaar 1977; Azzaroli 1982). Thus pigmy hippos, elephants, deer, and the bovid *Myotragus* occur in Rhodes, Cyprus, Crete, Sicily, Malta, Sardinia, and Mallorca and large rodents and insectivores in Sardinia, Corsica, the Balearics, and other islands.

Dwarfed Pleistocene elephants have also been found in various Indonesian islands, including Flores, Java, Sulawesi, Timor, and Sumba (Roth 1992). Additionally, dwarf mammoths have been reported from the Holocene of Wrangel Island in the Siberian Arctic (Vartanyan *et al.*, 1993). It has generally been thought that mammoths became extinct everywhere by about 9500 BP. This new discovery forces a re-evaluation of this view, because numerous teeth of dwarf mammoths have been dated as 7000–4000 BP. The island is thought to have been separated from the Siberian mainland at about 12 000 BP, and the mammoths are thought by the authors to have survived by escaping the climatic change on the continent that led to earlier extinction there.

MARINE BIOGEOGRAPHY

It has already been indicated that the establishment of land bridges between North and South America, Africa and Eurasia, and Eurasia and North America allowed free migration of mammals, but the convergence of land faunas necessarily correlates with divergence of marine faunas, and the converse—the principle of complementarity. Some of the relevant evidence from marine invertebrates is reviewed below, after which there is a brief discussion about features of biogeographic interest in the Pacific Ocean.

Effect of uplift of the Panama Isthmus and closure of Tethys

The existence of numerous cognate invertebrate species on the Pacific and Caribbean sides of the Panama Isthmus indicates a close evolutionary relationship, with isolation and divergence dating from the recent geological past (Vermeij 1978). The best evidence for dating the time of emergence of the isthmus comes from microfossils. The Panama and Colombia Basins, on the two sides of the isthmus, share two important Early Pliocene features among the planktonic foraminifers, the great abundance of sinistral coiling *Neogloboquadrina pachyderma* and a sinistral-to-dextral change in coiling direction preference in *Pulleniatina* at 3.5 Ma in both locations. Between 3.5 and 3.1 Ma *Pulleniatina* disappeared from the Colombia Basin, and this date is therefore the one preferred by Keigwin (1978) for the separation. The emergence of the isthmus need not necessarily have involved tectonic uplift because Keigwin notes that 3.1 Ma is close in time to the first buildup of ice in the northern hemisphere, implying a glacioeustatic sea-level fall.

Fossil macroinvertebrate faunas, especially molluscs, yield anomalously old ages compared with planktonic foraminifers, which is probably due to Neogene mass extinctions in the western Atlantic shelf faunas causing them to appear older (Jones and Hasson 1985). It is most reasonable to infer that elevation of the Panama Isthmus proceeded gradually, and sedimentological and palaeoceanographic studies suggest partial uplift and disruption of former current patterns as early as Late Miocene (Duque-Caro 1990). Total emergence of the entire isthmus probably did not occur until the latest Pliocene or earliest Pleistocene. This pattern of change is supported by research on strombinid gastropods (Jackson *et al.* 1993). There were only a small proportion of transisthmian species and substantial oceanic divergence of subgenera after the Late Miocene. However, the occurrence of two extremely similar pairs of exclusively Late Pliocene to modern species on opposite sides of the isthmus suggests that some interchange may still have been possible during the Late Pliocene.

With regard to the closure of Tethys in the Mediterranean region, the record of benthic Foraminifera indicates that in Early Miocene (Burdigalian) times the Mediterranean faunas became sharply differentiated from those of the Indo-Pacific and from this time on the emerging Mediterranean had no further contact with the latter (Adams 1967, 1983). A similar pattern of change is recognizable from the molluscan record (Hallam 1967*a*). At about the same time Atlantic and Mediterranean foraminifer faunas diverged (Berggren and Phillips 1969), presumably as a consequence of the suturing of North Africa with Spain.

The Bering Strait

The earliest suggestion of Cenozoic interoceanic migration through the Bering Strait is in the Late Miocene, according to Durham and MacNeil (1967). Over 125 species, mainly bivalves and gastropods, entered the Arctic–Atlantic region from the Pacific, but no more than 16 species of North Pacific Late Cenozoic faunas are of Atlantic origin. Migration through the Bering Strait continued, mostly from the Pacific, and is well documented in the Pliocene and Pleistocene, though the precise dating has not been well established. It need not have been more than intermittent.

Vermeij (1991) has reinvestigated the phenomenon, concentrating on how best to interpret the pronounced asymmetry, with invaders from the Pacific outnumbering those from the Atlantic by a ratio of 8 to 1 or more. He dates the main submergence and migration episode as Mid Pliocene (about 3.5 Ma). Of several hypotheses considered, evidence is found to support the one explaining the asymmetry as a consequence of extinctions being more severe in the Atlantic, allowing invading biota to gain a foothold there more readily.

The Pacific Ocean

Two interesting biogeographic studies of plankton have thrown light on palaeoceanographic changes in the Pacific.

The biogeographic patterns of planktonic Foraminifera were quantitatively mapped by Kennett *et al.* (1985) for three time intervals in the Miocene. During the Early Miocene the assemblages were dominated by different taxa in the east and west, but the assemblages became more similar late in the epoch. This change is interpreted firstly to reflect the development during the Mid Miocene of an equatorial undercurrent system when the Indonesian Seaway closed, and secondly the general intensification of tropical and gyral surface-water circulation, including a strengthening of the equatorial countercurrent, that resulted from steeper latitudinal gradients. In the Early Miocene either generally warmer surface waters or the presence of a deeper thermocline in the west favoured shallow-water dwellers such as *Globeriginoides*, while a shallower thermocline in the east favoured slightly deeper-dwelling forms. The differences were maintained as the equatorial countercurrent was weak. A separate equatorially distributed assemblage developed by the Late Miocene perhaps in response to the development of an equatorial undercurrent system and the strengthening of the equatorial countercurrent.

Romine (1985) undertook a factor analysis of 44 Late Miocene radiolarian species, which led to the distinction of three biogeographic provinces and depth-related categories:

(1) deep (subarctic assemblage)

(2) intermediate (transitional assemblage)

(3) shallow (tropical assemblage).

Several phenomena, such as the absence of an eastern Pacific tropical assemblage, the splitting up of a transequatorial assemblage into modern south-western and north-western tropical assemblages and the movement along the California coast of an eastern subpolar assemblage, strongly suggest major changes in water-mass structure in the equatorial Pacific. The likeliest cause is thought to be the emergence of the Panama Isthmus as a barrier to Atlantic–Pacific surface-water accumulation.

The flow of water from the Atlantic into the equatorial Pacific probably influenced the flow pattern of the equatorial countercurrent, causing some of the warm water that normally returns to the eastern tropical Pacific in the countercurrent to be diverted into the north-flowing western limb of the subtropical gyre. This led to warmer sea-surface temperatures at higher latitudes. As the flow from the Atlantic diminished through the latest Miocene and Early Pliocene, Pacific subarctic regions became progressively cooler. Gradual cessation of Atlantic–Pacific surface-water communication accelerated the global cooling trend initiated in the southern hemisphere. The termination of flow in the Mid Pliocene (Keigwin 1978) set the scene for the Quaternary world of ice-volume fluctuations.

Turning to the benthos, there is a long-standing interest in why Indo–Pacific diversity is higher than in the Atlantic–Caribbean region. To account for this phenomenon in reef corals Rosen (1984) has put forward an island clustering model that may have been relevant also to other benthic invertebrates. In Rosen's opinion, a simple application of the species – area effect is inadequate to explain the phenomenon. Although the Indo – Pacific region is larger, it is not easy to translate this into greater environmental heterogeneity. Furthermore the differences between the two regions cannot easily be attributed to climatic or other ecological factors.

A dispersalist model requires small founding populations to establish themselves beyond a pre-existing biogeographic barrier, whereas a vicariance model requires the emergence of a new biogeographic barrier. With regard to a series of islands as are found in the West Pacific, the closer sets of islands are more likely to fall within the critical dispersion distance for larvae. Glacioeustatically induced rises and falls of sea level would have led to successive events of allopatric speciation giving rise to new species, a kind of diversity pump analogous to that proposed by Valentine (1967) based on climatic changes. A sea-level lowering of about 100 m would create a landmass in Indonesia but a deep marine barrier would persist between there and Australia–New Guinea. In support of this model, there are many young coral genera in the Indo-Pacific region. Reasons why the Atlantic–Caribbean region has lower diversities could be because there has been more cold climate-induced extinction and because the eustatic diversity pump has been less effective. The species–area effect could have played an additional role.

THE INFLUENCE OF CLIMATE

Climate is the biogeographic leitmotiv of the Neogene, and especially the Quaternary. Even those biogeographic phenomena directly connected with changing continental and oceanic relationships are more often the consequence of climate-related glacioeustatic changes of sea level than plate tectonics. Climatic changes can be monitored by expanding and contracting ranges of both terrestrial and marine high- and low-latitude organisms.

The continents

With regard to high latitudes in the northern hemisphere, broad-leaved deciduous forests persisted into the Early Miocene both in Siberia and the Canadian Arctic. On the basis of leaf physiognomy, Wolfe (1980) estimates mean annual temperatures of 10–13°C at 55–60°N and 3–5°C at 70°N. By late Mid Miocene times the dominant vegetation was coniferous forest, and no tundra became established until the Late Neogene. In the southern hemisphere the Antarctic icecap dominated that continent from Mid Miocene times onward but rainforest persisted along water courses in Australia in the Miocene, with more open grassland in the centre of the country (Kemp 1978).

A good floral record has been preserved in Africa. There were aridity-adapted forests and savannas in East Africa and mixed savanna–woodland in the Saharan region in the Early and Mid Miocene (Axelrod and Raven 1978). By Late Miocene times cooler conditions had a widespread aridifying influence in southern Africa (Van Zinderen Bakker and Mercer 1986). Savanna as well as lowland rainforests and montane vegetation are thought to have occupied tropical Africa since the Miocene. Miocene times saw the initiation of fragmentation of closed forest, progressively to woodland and savanna in the Late Miocene and Early Pliocene. The significant contraction of tropical lowland rainforest and the establishment of savanna seems to fall in the Mid Miocene (Yemane *et al.* 1985). From palynological analysis, Yemane and his colleagues inferred post-depositional uplift of around 1000 m in the Ethiopian Highlands.

Extensive regional uplift is also inferred by Axelrod and Bailey (1976) from a study of Tertiary floras in part of the United States Western Interior. Mid Tertiary floras represent a temporal progression from the upper part of the mixed subtropical forest, to the upper margin of sclerophyll vegetation close to mixed conifer forest, to subalpine conifer forest. The implied decrease in mean annual temperature is attributed to the construction of volcanic piles and doming uplift in addition to a global cooling trend. The latest Oligocene and Mid Miocene floras yield evidence of subsequent epeirogenic uplift of about 1200 m, in an area which had been close to sea level in the Palaeocene. Since plateau and orogenic belt uplift was a major phenomenon globally, the role of increasing altitude in altering vegetation, in addition to climatic cooling, needs to be more fully evaluated in all the continents.

A subject of considerable biogeographic interest is the origin of exceptionally high biodiversity in tropical rainforests, most notably that of the Amazon. For a long time the notion has existed that such diversity was a product of enduring climatic stability. This belief was first challenged by Haffer (1969) who, to account for bird distributions, made the case for a series of Pleistocene refugia. He argued that the high diversity was the product of spatial heterogeneity and a series of expansions and contractions of forests related to climatic episodes. During glacial periods the rainforest shrank into isolated fragments and during interglacials the fragments fused together. Allopatric speciation took place during the first phase. When the ranges expanded the species were sufficiently distinct that they formed contact zones but retained their individuality. Haffer's hypothesis has proved popular and has been widely applied to other groups, including plants and butterflies (Whitmore and Prance 1987). It has also been held to be relevant to tropical rainforests in Africa (Van Zinderen Bakker and Mercer 1986; Van Zinderen Bakker and Coetzee 1988). There is no evidence, yet, that South-east Asian rainforests were ever reduced to isolated islands in a large savanna 'sea', although the lowland rainforest in Malaysia was, as elsewhere, of reduced extent during glacial maximum arid phases (Whitmore 1981*b*).

The idea of Pleistocene refugia even in the Amazon Basin remains controversial. Critics include population biologists, who argue that a combination of sympatric, parapatric, and stasipatric species differentiation in complex, heterogeneous environments could give rise to the observed patterns. Indeed, zones of environmental conformity and rapid transitions superimposed on endemic centres and contact zones show about as good a fit to distributional data as do refugia, so that more than one type of explanatory model may be needed (Brown 1987). In Lynch's (1988) view, it is not clear what sorts of data would be sufficient to falsify the refugia hypothesis.

According to Haffer (1987), studies of Quaternary sediments and their pollen content, together with landform analysis, reveal dramatic and wide-ranging climatic fluctuations which probably caused vast vegetational changes in the tropical lowlands of South America within the last million years. Geological, mineralogical, and micropalaeontological analyses of cores in the Caribbean and off north-eastern South America indicate climatic aridity during glacial phases in adjacent onshore regions. Certain soil formations, dissected and gullied land surfaces, as well as crusts found in parts of Amazonia indicate the previous existence of sparse vegetation pointing to at least one earlier period of drier climate than today.

As Haffer admits, there are relatively few data from the Amazon Basin itself (Colinvaux 1989). In Colinvaux's opinion it seems increasingly unlikely that the Amazon lowlands could have been sufficiently arid for the rain forests to have been destroyed. The observations most often cited to support aridity are the irregular layers of pebbles in laterite regoliths. These have been interpreted as pebbles washed out of a desert land surface, but they are more likely to be concretions. Furthermore, patterns of local endemism among various taxa are not yet well established, and could be cases of sampling failure. What is well established is the descent of altiplano vegetation in the Andes, indicating a 6–9°C temperature fall, with

lessened Andean precipitation at 20–12 ka BP. Coolness is well demonstrated, aridity less convincingly so. Significantly different forest communities are represented in pollen records for different parts of Amazonia
(Colinvaux *et al.* 1988). The authors suggest that areal differences in
modern Amazonian vegetation, rather than supposed patterns of refugia
in the past, account for disjunct distributions of Amazonian biota.

Although pollen analysis has been the most widely practised technique
for studying Quaternary climatic fluctuations some impressive results have
also been obtained from the analysis of beetle wing cases, which have a high
preservation potential. Although it might be expected that such a speciose
group would have experienced exceptionally high speciation rates in the
strongly changing environments of the Quaternary the evidence points
instead to significant morphological stability. An intimate relationship
between climatic changes and beetle distribution is well documented in
north-western Europe for the last 120 ka (Coope 1979).

The dung beetle *Aphodius holdereri* is an exceptionally interesting
illustrative case. At present it is restricted to the Tibetan Plateau above
3000 m. In the absence of a fossil record it could have been construed
as a Tibetan endemic adapted to high altitude environments, but the
fossil record indicates otherwise, because it is abundant in British fossil
assemblages accumulated close to sea level that date from the middle of
the last glaciation, accompanied by other species with varied ecological
preferences that are today characteristic of high cold steppes in Central
Asia. There is no need therefore to invoke any major differences in the
adaptation of fossil and living forms.

The fact that fossil assemblages even as far back as the Miocene strongly
resemble living communities suggests that ecological requirements have not
changed with time. If physiological evolution had been changing then there
should have been increasing distortion with age, unless (which appears
extremely unlikely) physiological change took place in the same degree and
direction independently. Coope points out the paradox that extraordinary
species stability coincides with numerous large-scale fluctuations of climate,
whereas one might expect high speciation and extinction rates. Coope's
explanation is that species tracked the environment rather than adapting or
going extinct. The migrations must have broken down the geographic barriers that separated the populations, thereby permitting genetic mixing.

Mammalian evolution and extinction.

From the Late Eocene to the present, the continents have undergone more
or less progressive net cooling, aridity, and seasonality, as demonstrated by
patterns of vegetational zonation. The vegetational zonation seen today,
ranging from tundra and taiga at high latitudes to tropical forest at the
equator, and with a significant proportion of low-latitude landmasses
desert, probably represents a more heterogeneous global vegetation than

at any other time in tetrapod history. It contrasts markedly with the floral evidence from the Palaeocene and Early Eocene of widespread closed-canopy tropical and subtropical forests. Janis and Damuth (1990) believe that climatic change in the Cenozoic has driven much of mammalian evolution and diversification.

Placentals have higher metabolic rates than marsupials, which gives them a competitive advantage in certain niches, for example where food resources permit increased metabolic rates (meat, seeds, and grass compared with invertebrates, fruit, and leaves), at small body masses and in very cold climates. As the climate became colder and grass became more available, placentals would have been favoured. The displacement in the Miocene of perissodactyls by artiodactyls as the dominant large herbivorous ungulates has traditionally been attributed to the superiority of the foregut system of fermentation in ruminant artiodactyls. But the patterns of radiation and extinction of the two orders do not fit the required inverse relationship. Instead, the replacement took place in the context of new resource types, to which artiodactyls adapted more readily than perissodactyls. Patterns of climatic change and concomitant vegetational structure and resource quality and abundance can explain the diversity reduction of perissodactyls, which are better adapted than artiodactyls for feeding on high diet, tropical non-deciduous foliage. Perissodactyls that 'escaped' this constraint until the present are either adapted to a specialized high-fibre grazing diet (horses) or have grown very large (rhinos) (Janis 1989).

Three important evolutionary trends in Cenozoic mammals are attributed by Janis and Damuth (1990) to extrinsic factors ultimately related to climatic change:

1. *Large body size in herbivores.* It is not possible at small size to subsist on a fibrous diet such as grass. Size change correlates well with habitat change.

2. *Hypsodonty.* High crowned teeth are found in herbivores that eat abrasive grass. Grasslands spread in area significantly in the Miocene. However, even browsers and mixed grazers-browsers have more hypsodont teeth if they feed in open habitats.

3. *Cursoriality.* The elongation of distal limb segments, loss of side toes, change in foot posture and limb motion are seen predominantly in carnivores and ungulates. There is no fossil evidence of coevolutionary coupling of predator and prey because ungulates acquired long limbs in the Early Miocene but no truly cursorial carnivores evolved until the Pliocene

and Pleistocene. The origin of cursoriality is related to the expansion of home-range area in more open habitats.

A case has also been made for ultimate climatic control of the evolution of our own genus (Stanley 1992). The appearance of large-brained early *Homo* at about 2.4 Ma followed closely on global climatic changes about 2.5 Ma that signalled the onset of the modern ice age. In Africa forests shrank and grassy habitats expanded. This is the type of change that could have forced some australopithecine populations to abandon arboreal activities, and a single fully terrestrial population could have evolved into *Homo*. Encephalization would have followed selection pressures favouring cleverness, tool use, and advanced socialization, promoting predator avoidance and expansion of trophic resources. Probably the only morphogenetic mechanism by which such dramatic encephalization could have evolved was extension of high foetal growth rates into the post-natal period, and this could not have happened until human ancestors had become fully terrestrial.

Webb and Barnosky (1989) recognize six major episodes of Neogene mammal extinction in North America. The largest, involving the loss of 62 genera, was in the Late Miocene (Mid Hemphillian – about 6 Ma). The next largest is dated as Late Rancholabrean – about 10 ka, with 43 genera lost, and the third largest in the Late Pliocene – about 1.9 Ma, with 35 genera lost. The Neogene mammal extinctions all correlate with intervals of rapid climatic change and decreasing equability, and all but the youngest are generally accepted as being climatically induced. This youngest event has been attributed to overkill by early Man (Martin 1984) but Webb and Barnosky contest this hypothesis. They point out for instance that only five of the 37 large vertebrate species that disappeared have been documented as butchered or killed by our own species and while bison were undoubtedly killed they survived the extinction event. Capybaras, tapirs, llamas, peccaries, and jaguars were lost from temperate northern latitudes but survive in the neotropics. Other Late Pleistocene casualties in North America, such as horses, camels, elephants, and cheetahs, survive in Old World refugia. It is perhaps significant that Africa, where the mammalian megafauna has largely survived, is the one continent where extensive tropical savannas survived the Pleistocene climatic changes.

The oceans

Although the general trend in the Neogene was towards a cooler global climate there were many oscillations on a range of scales. There was for example an early warming phase in the Latest Oligocene–Early Miocene, marked by a radiation of larger benthic foraminifers such as *Lepidocyclina* and *Miogypsina*, and an associated spread of shallow-water carbonate facies in the Mediterranean region. A marine transgression in the Burdigalian corresponds with a remarkable radiation of giant molluscs (Rögl and Steiniger 1984).

As noted earlier, the evidence of oxygen isotopes indicates a significant growth of the Antarctic icecap in Mid Miocene times. There is a corresponding expansion in the Indo–Pacific of polar-subpolar planktonic foraminiferal provinces and a latitudinal contraction of tropical provinces (Kennett *et al.* 1985). Using the same factor analysis technique as used for the Palaeogene (Haq *et al.* 1977) Haq (1980) has been able to recognize four Miocene warming–cooling cycles in both hemispheres for calcareous nannoplankton. There is good agreement between these cycles based on the expansion and contraction of the ranges of high- and low-latitude taxa, and oxygen isotope data.

For the Quaternary deep sea cores containing planktonic Foraminifera belonging to the same species as living forms whose temperature tolerances are known have been used. Thus *Globigerina pachyderma* is a species now dominant in surface sediments beneath cold Arctic waters characterized by seasonal sea-ice cover, and is dominant during glacial isotope stages as far south as 50°N in the Atlantic. At the other extreme, there is a group indicative of the warmer waters of the North Atlantic drift, e.g. *Globorotalia inflata*, and still warmer and more saline waters of the northern subtropical gyre, e.g. *Globeriginoides ruber*. These species are most abundant in southern cores in the North Atlantic but range in lesser percentages far to the north during the maximum interglaciations (Ruddiman and McIntyre 1984). Winter and summer sea-surface temperatures can be estimated using transfer functions (Imbrie and Kipp 1971). The influence of orbital forcing of climate (Milankovitch cycles) is evident from Ruddiman and McIntyre's analysis. The 100 ka cycle is strong everywhere and dominant in most of their core records. The 23 ka cycle is very strong near 40°N and the 41 ka cycle strongest at 50–55°N.

The Neogene calcareous planktonic microfossils of the Southern Ocean adjacent to the Antarctic are consistently of low diversity and high dominance, but subantarctic assemblages much more diverse. The Neogene marked the time when siliceous microfossils, notably diatoms, became dominant biogenic elements. In the latest Miocene there was a northward movement of the Antarctic convergence and water mass and hence the siliceous biogenic province. The Quaternary experienced a maximum northward extension of Antarctic waters and increased productivity of siliceous plankton (Kennett 1978, 1980).

Turning to the neritic benthos, Vermeij (1986) has indicated that the seas off West Africa and in the tropical East Pacific served as a Quaternary refuge for warm-water molluscs whose Miocene and Pliocene distributions extended over a wider range of latitude. Several bivalve and gastropod species occur in the shallow-water temperate biotas off Europe and southern Africa, but not off tropical West Africa. Studies of the fossil record show that these trans-equatorial distributions were achieved before the Late Pliocene rather than during glacial episodes of the Pleistocene, when the tropical belt might have been narrower. The likeliest time of migration is Mid Pliocene, the time of an important migration from the North Pacific into the Atlantic. The West African tropical zone must have remained wide enough or warm enough from the Late Pliocene onwards to have acted as an effective barrier even during glacial episodes (Vermeij 1992).

Vermeij's (1992) study raises the subject of antitropicality, which is also considered by Lindberg (1991), who analysed the distribution of some East Pacific molluscs and endeavoured to demonstrate that the use of fossils in studying this phenomenon is vital. Lindberg recognizes at least two major events:

1. Pliocene : the appearance of *Chama, Crenomytilus, Cryptomya*, and *Tegula* (*Chlorotoma*) in the southern hemisphere and *Discocurria* in the northern hemisphere. hemisphere.

2. Early Pleistocene : appearance of *Fusitriton* and *Argopecten* off Chile and *Fissurella* and *Fissurellidea* off California.

The diversity of taxa, habitats, and life history strategies, together with the bidirectionality of timing, argues against a single cause. Some taxa such as *Fusitriton* and *Argobuccinum* probably crossed the equatorial zone by submergence, but this was not possible for the kelp *Macrocystis*, which requires surface waters. According to Lindberg the fossil record of many antitropical taxa indicates that they first appeared in higher latitudes, and therefore the scenario of extinction of an ancestral equatorial population and formation of two higher latitude descendents is not supported. Therefore some form of dispersal across the tropics seems the most likely explanation.

The multiplication of Cenozoic marine provinces consequent upon steepening of latitudinal thermal gradients should enhance speciation (Valentine 1973). It has been argued that the global species diversities of shallow seas have increased by a factor of three or so in the Neogene alone (Valentine *et al.* 1978). Valentine (1984) has more recently maintained that the prevalent pattern for Neogene shelled molluscs is a dual radiation to produce distinct high-and low- latitude clades.

Additionally, typically tropical groups such as strombids and tridacnids have had significant Neogene radiations. Valentine considers that the patterns fit a double climatic diversity-pump model, giving opportunities for diversification in both low and high latitudes. The hypothesis could be supported if it could be demonstrated independently that there was a Neogene temperature rise in the tropics, with Early Neogene temperatures there being several degrees cooler than today.

This is precisely what has been claimed for surface waters by Shackleton (1984) on the basis of oxygen isotope analyses of planktonic Foraminifera. He has argued for stable temperatures of around 18°C through the Palaeogene to the Late Miocene, with modern temperatures of 28°C being geologically anomalous. Shackleton's interpretation has been strongly challenged, by Adams *et al.* (1990) on the basis of the distribution of three unrelated and diverse groups. At present their distribution is more or less directly controlled by sea-surface temperatures and at least two, the larger foraminifers and zooxanthellate corals, have a good fossil record, while that of mangroves, known from pollen, fruits, and rhizoliths, is moderate. As regards the corals, palaeosynecological evidence suggests a latitudinal belt about 10° wider at times from the Eocene to the Mid Miocene than today. The symbiont-bearing larger foraminifers are bounded by the 18–20°C isotherm for the warmest month and are confined, like the corals, to the euphotic zone. They are restricted to the tropics except where warm currents such as the Gulf Stream exist. Their latitudinal distribution and diversity was greater in Mid Eocene and Early Miocene times than today. Shackleton may have been misled by his assumption of an ice-free world prior to the Mid Miocene, which leads to the calculation of surprisingly cool tropical sea-surface temperatures in the Oligocene (because oxygen isotope composition of seawater is influenced by changing ice volume as well as by temperature). More recent work suggests the existence of substantial amounts of Antarctic ice at least from earliest Oligocene times onwards (Kennett and Barker 1990).

Mass extinction appears not to have been a major phenomenon among the marine biota, but Stanley (1984, 1987) makes a good case for such an event among Late Pliocene molluscs. However, this was a regional not a global event, related to falling temperatures in the western North Atlantic, associated with growth of the Canadian, Greenland, and Scandinavian ice-caps. As temperature fell species migrated southwards but were restricted by the presence of Florida and the Gulf of Mexico. On the Pacific margins, there was no such restriction to movement towards the tropics and few indications of such a sharp temperature fall, and no extinction event has been recorded here.

According to Jackson *et al.* (1993) the mass extinction event first documented by Stanley for molluscs north of the Caribbean occurred in a vastly greater region around the Caribbean and affected corals as well as molluscs. They consider that changes in patterns of upwelling and nutrient distribution may have been more important factors in causing the mass extinction than refrigeration due to the onset of the Pleistocene glaciation, because there was no evident temperature decline in the southern Caribbean after the Pliocene and only a slight decline in Florida.

APPENDIX

The Phanerozoic Time Scale
(simplified from Harland *et al.* 1989)

Era	Period	Epoch	Stage	Age (Ma)
Cenozoic	Neogene	Holocene		0.01
		Pleistocene		1.64
		Pliocene	Piacenzian	
			Zanclian	5.2
		Miocene	Messinian	
			Tortonian	
			Serravalian	
			Langhlian	
			Burdigalian	
			Aquitanian	23.3
	Palaeogene	Oliogocene	Chattian	
			Rupelian	35.4
		Eocene	Priabonian	
			Bartonian	
			Lutetian	
			Yypresian	56.5
		Palaeocene	Thanetian	
			Danian	65.0
Mesozoic	Cretaceous		Maastrictian	
			Campanian	
			Santonian	
			Coniacian	
			Turonian	
			Cenomanian	
			Albian	
			Aptian	
			Barremian	
			Hauterivian	
			Valanginian	
			Berriasian	145.6
	Jurassic		Tithonian	
			Kimmeridgian	
			Oxfordian	
			Callovian	

Era	Period	Epoch	Stage	Age (Ma)
			Bathonian Bajocian Aalemian Toarcian Pliensbachian Sinemurian Hettangian	
				208.0
	Triassic		Rhaetian Norian Carnian Ladinian Anisian Spathian Nammalian Griesbachian	
				245.0
Upper Palaeozoic	Permian		Changxingian Longtanian Capitanian Wordian Ufimian Kungurian Artinskian Sakmarian Asselian	
				290.0
	Carboniferous	Stephanian Westphalian Namurian Visean Tournaisian		
				362.5
	Devonian		Famennian Frasnian Givetian Eifelian Emsian Pragian Lochkovian	
				408.5
	Silurian	Pridoli Ludlow Wenlock Llandovery		
				439.0

Era	Period	Epoch	Stage	Age (Ma)
Lower Palaeozoic	Ordovician	Ashgill Caradoc Llandeilo Llanvirn Arenig Tremadoc		
				510.0
	Cambrian	Merioneth St David's Caerfai		
				570.0

Bibliography

Abele, L.G. and Walters, K. (1979). Marine benthic diversity: a critique and alternative explanation. *Journal of Biogeography*, **6**, 115–26.

Adams, C.G. (1967). Tertiary Foraminifera in the Tethyan, American, and Indo-Pacific provinces. In *Aspects of Tethyan biogeography* (ed. C.G. Adams and D.V. Ager), pp. 195–217. Systematics Association Publication no. 7.

Adams, C.G. (1973). Some Tertiary Foraminifera. In *Atlas of palaeobiogeography* (ed. A. Hallam), pp. 453–68. Elsevier, Amsterdam.

Adams, C.G. (1983). Speciation, phylogenesis, tectonism, climate and eustasy: factors in the evolution of Cenozoic larger foraminiferal bioprovinces. In *Evolution, time and space: the emergence of the biosphere* (ed. R.W. Sims, J.H. Price, and P.E.S. Whalley), pp. 255–98. Academic Press, London.

Adams, C.G., Lee, D.E., and Rosen, B.R. (1990). Conflicting isotopic and biotic evidence for tropical sea-surface temperatures during the Tertiary. *Palaeogeography, Palaeoclimatology, Palaeoecology*, **77**, 289–313.

Ager, D.V. (1973). Mesozoic brachiopods. In *Atlas of palaeobiogeography* (ed. A. Hallam), pp. 431–36. Elsevier, Amsterdam.

Ager, D.V. and Walley C.D.(1977). Mesozoic brachiopod migrations and the opening of the North Atlantic. *Palaeogeography, Palaeoclimatology, Palaeoecology*, **21**, 85–99.

Aiello, L.C. (1993). The origin of New World monkeys. In *The Africa–South America connection* (ed. W. George and R. Lavocat), pp. 100–18. Clarendon Press, Oxford.

Aita, Y. and Spörli, K.B. (1992). Tectonics and paleobiogeographic significance of radiolarian microfaunas in the Permian to Mesozoic basement rocks of the North Island, New Zealand. *Palaeogeography, Palaeoclimatology, Palaeoecology* **96**, 103–25.

Albino, A.M. (1993). Snakes from the Paleocene and Eocene of Patagonia (Argentina): paleoecology and coevolution with mammals. *Historical Biology* **7**, 51–69.

Allison, P.A. and Briggs, D.E.G. (1993). Paleolatitudinal sampling bias, Phanerozoic species diversity, and the end-Permian extinction. *Geology*, **21**, 65–8.

Arkell, W.J. (1956). *Jurassic geology of the world*. Oliver and Boyd, Edinburgh.

Aubry, M.-P. (1983). Late Eocene to early Oligocene calcareous nannoplankton biostratigraphy and biogeography. *Bulletin of the American Association of Petroleum Geologists*, **67**, 415.

Audley-Charles, M.G. (1981). Geological history of the region of Wallace's line. In *Wallace's line and plate tectonics* (ed. T.C. Whitmore), pp. 24–35. Clarendon Press, Oxford.

Audley-Charles, M.G. (1983). Reconstruction of eastern Gondwanaland. *Nature*, **306**, 48–50.

Audley-Charles, M.G., Ballantyne, P.D., and Hall, R. (1988). Mesozoic–Cenozoic rift-drift sequence of Asian fragments from Gondwanaland. *Tectonophysics*, **155**, 317–30.

Axelrod, D.I. (1970). Mesozoic paleogeography and early angiosperm history. *Botanical Review*, **36**, 277–319.

Axelrod, D.I. (1984). An interpretation of Cretaceous and Tertiary biota in polar regions. *Palaeogeography, Palaeoclimatology, Palaeoecology*, **45**, 105–47.

Axelrod, D.I. and Bailey, H.P. (1976). Tertiary vegetation, climate and altitude of the Rio Grande depression, New Mexico–Colorado. *Paleobiology*, **2**, 235–54.

Axelrod, D.I. and Raven, P.H. (1978). Late Cretaceous and Teritary vegetation history of Africa. In *Biogeography and ecology of Southern Africa* (ed. M.J.A Werger), pp. 77–130. Junk, The Hague.

Azzaroli, A. (1982). Insularity and its effects on terrestrial vertebrates: evolutionary and biogeographic aspects. In *Palaeontology, essential of historical geology* (ed. E. Montanaro Gallitelli), pp. 193–213. S.T.E.M. Mucchi Modena Press.

Azzaroli, A. (1983). Quaternary mammals and the 'end-Villafranchian' dispersal event – a turning point in the history of Eurasia. *Palaeogeography, Palaeoclimatology, Palaeoecology*, **44**, 117–39.

Azzaroli, A. and Guazzone, G. (1979). Terrestrial mammals and land connections in the Mediterranean before and during the Messinian. *Palaeogeography, Palaeoclimatology, Palaeoecology*, **29**, 155–67.

Bakker, R.T. (1977). Tetrapod mass extinctions – a model of the regulation of speciation rates and immigration by cycles of topographic diversity. In *Patterns of evolution as illustrated by the fossil record* (ed. A. Hallam), pp. 439–68. Elsevier, Amsterdam.

Ball, I.R. (1976). Nature and formulation of biogeographic hypotheses. *Systematic Zoology*, **24**, 407–30.

Ball, I.R. (1983). Planarians, plurality and biogeographical explanations. In *Evolution, time and space: the emergence of the biosphere* (ed. R.W. Sims, J.H. Price and P.E.S. Whalley), pp. 409–30. Academic Press, London.

Bambach, R.K. (1977). Species richness in marine benthic habitats through the Phanerozoic. *Paleobiology*, **3**, 152–67.

Bambach, R.K. (1990). Late Palaeozoic provinciality in the Marine Realm. In *Palaeozoic, palaeogeography and biogeography, Geological Society Memoir no. 12. (ed. W.S. McKerrow and C.R McKerrow and C.R. Scotese)*, pp. 307–23.

Bambach, R.K. (1993). Seafood through time: Changes in biomass, energetics, and productivity in the marine ecosystem *Paleobiology* **19** 372–97.

Barnard, P.D.W. (1973). Mesozoic floras. *Special Papers in Palaeontology*, **12**, 175–87.

Barron, E.J. (1984). Climatic implications of variable obliquity explanation of Cretaceous-Paleogene high-latitude floras. *Geology*, **12**, 595–98.

Barry, J.C., Morgan, M.E., Winkler, A.J., Flynn, L.J., Lindsay, E.H., Jacobs, L.L., and Pilbeam, D. (1991). Faunal interchange and Miocene terrestrial vertebrates of southern Asia. *Paleobiology*, **17**, 231–45.

Beauvais, L. (1993). Corals of the circum-Pacific region. In *The Jurassic of the circum-Pacific* (ed. G.E.G. Westermann), pp. 324–7. Cambridge University Press, Cambridge.

Belasky, P. (1992). Assessment of sampling bias in biogeography by means of a probabilistic estimate of taxonomic diversity: application to modern Indo-Pacific reef corals. *Palaeogeography, Palaeoclimatology, Palaeoecology*, **99**, 243–70.

Belasky, P. and Runnegar, B. (1993). Biogeographic constraints for tectonic reconstructions of the Pacific region. *Geology*, **21**, 979–82.

Benson, R.H. (1975). The origin of the psychrosphere as recorded in changes of deep-sea ostracode assemblages. *Lethaia*, **8**, 69–83.

Benton, M.J. (1987*a*). Mass extinctions among families of non-marine tetrapods: the data. Mémoires de la Société géologique de la France, **150**, 21–32.

Benton, M.J. (1987*b*). Progress and competition in macroevolution. *Biological Reviews*, **62**, 305–38.

Benton, M.J. (1991). What really happened in the Late Triassic? *Historical Biology*, **5**, 263–78.

Berggren, W.A. and Phillips, J.D. (1969). Influence of continental drift on the distribution of Tertiary benthonic Foraminifera in the Caribbean and Mediterranean regions. *Contributions of the Woods Hole Oceanographic Institution* no. 2376.

Bergström, S.M. (1990). Relations between conodont provincialism and the changing palaeogeography during the Early Palaeozoic. In *Palaeozoic, palaeogeography and biogeography*, Geological Society Memoir no. 12, (ed. W.S. McKerrow and C.R. Scotese), pp. 105–21.

Berry, W.B.N. (1973). Silurian–Early Devonian graptolites. In *Atlas of palaeobiogeography* (ed. A. Hallam), pp. 81–7. Elsevier, Amsterdam.

Berry, W.B.N. and Wilde, P. (1990). Graptolite biogeography: implications for palaeogeography and palaeoceanography. In *Palaeozoic, palaeogeography and biogeography* Geological Society Memoir no. 12, (ed. W.S. McKerrow and C.R. Scotese), pp. 129–37.

Besse, J., Courtillot, V., Pozzi, J.P., Westphal, M., and Zhou, Y.X. (1984). Palaeomagnetic estimates of crustal shortening in the Himalayan thrusts and Zangbo suture. *Nature*, **311**, 621–6.

Blendinger, W., Furnish, W.M., and Glenister, B.F. (1992). Permian cephalopod limestones, Oman Mountains: evidence for a Permian seaway along the northern margin of Gondwana. *Palaeogeography, Palaeoclimatology, Palaeoecology*, **93**, 13–20.

Boersma, A., Premoli-Silva, I., and Shackleton, N.J. (1987). Atlantic Eocene planktonic foraminiferal paleohydrographic indicators and stable isotope paleohydrographic indicators and stable isotope paleoceanography. *Paleoceanography*, **2** 287–331.

Bonaparte, J.F. (1979). Dinosaurs: a Jurassic assemblage in Patagonia. *Science*, **205**, 1377–9.

Bonaparte, J.F. (1984). Late Cretaceous faunal interchanges of terrestrial vertebrates between the Americas. In *3rd Symposium on Mesozoic terrestrial ecosystems* (ed. W.E. Reif and F. Westphal), pp. 19–24. Attempto, Tübingen.

Boucot, A.J. (1975). *Evolution and extinction rate controls*. Elsevier, Amsterdam.

Boucot, A.J. (1990). Silurian biogeography. In *Palaeozoic, palaec ⟩graphy and biogeography*, Geological Society Memoir no. 12, (ed. W.S. McKerrow and C.R. Scotese), pp. 191–6.

Boucot, A.J. and Gray, J. (1983). A Paleozoic Pangaea. *Science*, **222**, 571–81.

Boucot, A.J. and Johnson, J.G. (1973). Silurian brachiopods. In *Atlas of palaeobiogeography* (ed. A. Hallam), pp. 59–65. Elsevier, Amsterdam.

Bown, T.M. and Simons, E.L. (1984). First record of marsupials (Metatheria: Polyprotodonta) from the Oligocene in Africa. *Nature*, **308**, 447–9.

Boyd, A. (1990). The Thyra Ø Flora: toward a understanding of the climate and vegetation during the Early Tertiary in the High Arctic. *Reviews of Palaeobotany and Palynology*, **62**, 189–203.

Brenner, G.J. (1976). Middle Cretaceous floral provinces and early migrations of angiosperms. In *Origin and early evolution of angiosperms*, pp. 23–47. Columbia University Press, New York.

Briggs, J.C. (1974). *Marine zoogeography*. McGraw-Hill, New York.

Briggs, J.C. (1987). *Biogeography and plate tectonics*. Elsevier, Amsterdam.

Brown, J.H. (1988). Species diversity. In *Analytical biogeography* (ed. A.A. Myers and P.S. Giller), pp. 57–89. Chapman and Hall, London.

Brown, K.S. (1987). Conclusions, synthesis and alternative hypotheses. In *Biogeography and Quaternary history in tropical America*, (ed. T.C. Whitmore and G.T. Prance), pp. 175– 212. Clarendon Press, Oxford.

Brundin, L.Z. (1988). Phylogenetic biogeography. In *Analytical biogeography* (ed. A.A. Myers and P.S. Giller), pp. 343–69. Chapman and Hall, London.

Brunn, A.F. (1957). Deep sea and abyssal depths. In *Treatise on marine ecology and paleoecology*. Geological Society of America Memoir 67, pp. 641–72.

Buffetaut, E. (1982). A ziphodont mesosuchian crocodile from the Eocene of Algeria and its implications for vertebrate dispersal. *Nature*, **300**, 176–8.

Buffetaut, E. (1985*a*). The palaeobiogeographic significance of the Mesozoic continental vertebrates from South-East Asia. In *Paléogéographie de l' Indie, du Tibet et du Sud-est Asiatique*. Memoires de la Société géologique de la France 147, pp. (ed. E. Buffetaut, J.J. Jaeger, and J.C. Rage). 37–42.

Buffetaut, E. (1985*b*). Présence de Trematochampside (Crocodylia, Mesosuchia) dans le Crétacé supérieur du Brésil. Implications paléogéographiques. *Comptes Rendus de l' Academie des Sciences, Paris*, Sér. II, **301**, 1221–4.

Buffetaut, E. and Rage, J.C. (1993). Fossil amphibians and reptiles and the Africa–South America connection. In *The Africa–South America connection* (ed. W. George and R. Lavocat), pp. 87–99. Clarendon Press, Oxford.

Buffon, G.L.L., Comte de (1761). *Histoire naturelle, generale et particulière*, vol. 9. Imprimerie Royale, Paris.

Burrett, C.F. (1973). Ordovician biogeography and continental drift. *Palaeogeography, Palaeoecology, Palaeoclimatology*, **13**, 161–201.

Burrett, C.F. (1974). Plate tectonics and the fusion of Asia. *Earth and Planetary Science Letters*, **21**, 181–9.

Bullard, E.C., Everett, J.E., and Smith, A.G. (1965). The fit of the continents around the Atlantic. *Philosophical Transactions of the Royal Society of London*, **258A**, 41–51.

Burrett, C., Long, J., and Stait, B. (1990). Early–Middle Palaeozoic biogeography of Asian terranes derived from Gondwana. In *Palaeozoic, palaeogeography and biogeography*, Geological society Memoir no. 12. (ed. W.S. McKerrow and C.R. Scotese), pp. 163–74.

Buzas, M.A., Koch, C.F., Culver, S.J., and Sohl, N.F. (1982). On the distribution of species occurrence. *Paleobiology*, **8**, 143–50.

Callomon, J.H. (1984). A review of the biostratigraphy of the post-Lower Bajocian Jurassic ammonites of Western and Northern North America. *Geological Society of Canada Special Paper* **27**, 143–74.

Cande, S.C. and Mutter, J.C. (1982). A revised identification of the oldest sea-floor spreading anomalies between Australia and Antarctica. *Earth and Planetary Science Letters*, **58**, 151–60.

Candolle, A.P. de (1820). *Geographie botanique*. In *Dictionnaire des sciences naturelles* Vol. 18, 359–422.

Caputo, M.V. and Crowell, J.C. (1985). Migration of glacial centres across

Gondwana during the Paleozoic era. *Bulletin of the Geological Society of America*, **96**, 1020–36.

Carey, S.W. (1976). *The expanding earth*. Elsevier, Amsterdam.

Cariou, E. (1973). Ammonites of the Callovian and Oxfordian. In *Atlas of palaeobiogeography* (ed. A. Hallam), pp. 287–95. Elsevier, Amsterdam.

Casey, R. and Rawson, P.F. (1973). *The boreal Lower Cretaceous*. Seel House Press, Liverpool.

Catalano, R., Di Stefano, P., and Kozur, H. (1991). Permian circumpacific deep-water faunas from the western Tethys (Sicily, Italy) – new evidence for the position of the Permian Tethys. *Palaeogeography, Palaeoclimatology, Palaeoecology*, **87**, 75–108.

Catalano, R., Di Stefano, P., and Kozur, H. (1991). Permian circumpacific deep-water faunas from the western Tethys (Sicily, Italy)—new evidence for the position of the Permian Tethys. *Palaeogeography, Palaeoclimatology, Palaeoecology*, **87**, 75–108.

Challinor, A.B., Doyle, P., Howlett, P.J., and Nal'nyaeva, T.I. (1992). In *The Jurassic of the circum-Pacific* (ed. G.E.G. Westermann), pp. 334–41. Cambridge University Press, Cambridge.

Chaloner, W.G. and Creber, G.T. (1988). Fossil plants as indicators of Late Palaeozoic plate positions. In *Gondwana and Tethys* (ed. M.G. Audley-Charles and A. Hallam), pp. 201–10. Oxford University Press, Oxford.

Chaloner, W.G. and Hallam, A. (1989). *Evolution and extinction*. The Royal Society, London.

Chaloner, W.G. and Lacey, W.S. (1973a). The distribution of Late Palaeozoic floras. *Special Papers in Palaeontology* no. 12, 271–89.

Chaloner, W.G. and Meyen, S.V. (1973b). Carboniferous and Permian floras of the northern continents. In *Atlas of palaeobiogeography* (ed. A. Hallam), pp. 169–86. Elsevier, Amsterdam.

Chaloner, W.G. and Sheerin, A. (1981). The evolution of reproductive strategies in early land plants. In *Evolution today* (ed. G.C.E. Scudder and G.L. Reveal), pp. 93–100. Hunt Institute of Botanical Documentation, Pittsburgh.

Chapronière, G.C.H. (1980). Influence of plate tectonics on the distribution of Late Palaeogene to Early Neogene larger foraminiferids in the Australasian region. *Palaeogeography, Palaeoclimatology, Palaeoecology*, **31**, 299–317.

Charig, A.J. (1973). Jurassic and Cretaceous dinosaurs. In *Atlas of palaeobiogeography* (ed. A. Hallam), pp. 339–52. Elsevier, Amsterdam.

Chateauneuf, J.-J. (1980). Palynostratigraphie et paléoclimatologie de l'Eocéne supérieur et de l'Oligoène du Bassin de Paris (France). *Mémoires de la Bureau de Recherches Géologiques et Minières* 116.

Chatterjee, S. (1992). A kinematic model for the evolution of the Indian plate since the Late Jurassic. In *New concepts in global tectonics* (ed. S. Chatterjee and N. Hotton), pp. 33–62. Texas Technical University Press, Lubbock.

Cheetham, A.H. and Hazel, J.E. (1969). Binary (presence–absence) similarity coefficients. *Journal of Paleontology*, **43**, 1130–6.

Churkin, M. (1972). Western boundary of the North American continental plate in Asia. *Bulletin of the Geological Society of America*, **83**, 1027–36.

Coates, A.G. (1973). Cretaceous Tethyan coral-rudist biogeography related to the evolution of the Atlantic Ocean. *Special Papers in Palaeontology*, **12**, 169–74.

Cocks, L.R.M. and Fortey, R.A. (1982). Faunal evidence for oceanic separations in the Palaeozoic of Britain. *Journal of the Geological Society*, **139**, 465–78.

Cocks, L.R.M. and Fortey, R.A. (1988). Lower Palaeozoic facies and faunas around Gondwana. In *Gondwana and Tethys* (ed. M.G. Audley-Charles and A. Hallam), pp. 183–200. Oxford University Press, Oxford.

Cocks, L.R.M. and Fortey, R.A. (1990). Biogeography of Ordovician and Silurian faunas. In *Palaeozoic palaeogeography and biogeography*, Geological Society Memoir no. 12.(ed. W.S. McKerrow and C.R. Scotese), pp. 97–104.

Cocks, L.R.M. and McKerrow, W.J. (1973). Brachiopod distributions and faunal provinces in the Silurian and Lower Devonian. *Special Papers in Palaeontology*, no. 12, 291–304.

Colinvaux, P.A. (1989). Ice-age Amazon revisited. *Nature*, **340**, 188–9.

Colinvaux, P.A., Frost, M., Frost, I., Kam-Biu, and C., Liu Steinitz-Kannan, (1988). Three pollen diagrams of forest disturbance in the western Amazon basin. *Reviews of Palaeobotany and Palynology*, **55**, 73–81.

Collinson, M.E., Fowler, K., and Boulter, M.C. (1981). Floristic changes indicate a cooling climate in the Eocene of southern England. *Nature*, **291**, 315–17.

Coney, P.J. (1979). Mesozoic–Cenozoic cordilleran plate tectonics. *Memoirs of the Geological Society of America*, **152**, 33–50.

Connor, E.F. and McCoy, E.D. (1979). The statistics and biology of the species – area relationship. *American Naturalist*, **113**, 791–833.

Coope, G.R. (1979). Late Cenozoic fossil Coleoptera: evolution, biogeography and ecology. *Annual Reviews of Ecology and Systematics*, **10**, 247–67.

Cooper, R.A., Fortey, R.A., and Lindholm, K. (1991). Latitudinal and depth zonation of early Ordovician graptolites. *Lethaia*, **24**, 199–218.

Copper, P. (1986). Frasnian/Famennian mass extinction and cold-water oceans. *Geology*, **14**, 835–9.

Coryndon, S.C. and Savage, R.J.G. (1973). The origin and affinities of African mammal faunas. *Special Papers in Palaeontology*, **12**, 121–35.

Cox, C.B. (1973). Triassic tetrapods. In *Atlas of palaeobiogeography* (ed. A. Hallam), pp. 213–23. Elsevier, Amsterdam.

Cox, C.B. (1974). Vertebrate palaeodistributional patterns and continental drift. *Journal of Biogeography*, **1**, 75–94.

Cox, C.B. (1990). New geological theories and old biogeographical problems. *Journal of Biogeography*, **17**, 117–30.

Cracraft, J. (1973). Continental drift, paleoclimatology, and the evolution and biogeography of birds. *Journal of Zoology*, **169**, 455–545.

Cracraft, J. (1980). Biogeographic patterns of terrestrial vertebrates in the southwest Pacific. *Palaeogeography, Palaeoclimatology, Palaeoecology*, **31**, 353–69.

Crame, J.A. (1986). Late Mesozoic bipolar bivalve faunas. *Geological Magazine*, **123**, 611–18.

Crame, J.A. (1992*a*). Evolutionary history of the polar regions. *Historical Biology*, **6**, 37–60.

Crame, J.A. (1992*b*). Late Cretaceous palaeoenvironments and biotas: an Antarctic perspective. *Antarctic Science*, **4**, 371–82.

Crame, J.A. (1993). Bipolar molluscs and their evolutionary implication. *Journal of Biogeography*, **20**, 145–61.

Craw, R. (1988). Panbiogeography: method and synthesis in biogeography. In *Analytical biogeography* (ed. A.A. Myers and P.S. Giller), pp. 405–35. Chapman and Hall, London.

Creber, G.T. and Chaloner, W.G. (1985). Tree growth in the Mesozoic and early Tertiary and the reconstruction of palaeoclimates. *Palaeogeography, Palaeoclimatology, Palaeoecology*, **52**, 35–60.

Crick, R.E. (1990). Cambro–Devonian biogeography of nautiloid cephalopods. In *Palaeozoic palaeogeography and biogeography*, Geological Society Memoir no. 12.(ed. W.S. McKerrow and C.R. Scotese), pp. 147–61.

Croizat, L. (1952). *Manual of phytogeography*. Junk, The Hague.

Croizat, L. (1958). *Panbiogeography*, vol. 1, 2a, 2b. Published by the author, Caracas.

Croizat, L. (1964). *Space, time, form: the biological synthesis*. Published by the author, Caracas.

Croizat, L., Nelson, G., and Rosen, D.E. (1974). Centers of origin and related concepts. *Systematic Zoology*, **23**, 265–87.

Dagys, A.S. (1993). Geographic differentiation of Triassic brachiopods. *Palaeogeography, Palaeoclimatology, Palaeoecology*, **100**, 79–87.

Dalziel, I.W.D. (1992). Antarctica; a tale of two supercontinents. *Annual Reviews of Earth and Planetary Sciences*, **20**, 501–26.

Dalziel, I.W.D. and Elliot, D.H. (1982). West Antarctica: problem child of Gondwanaland. *Tectonics*, **1**, 3–19.

Damborenea, S.E. (1993). Early Jurassic South American pectinaceans and circum-Pacific palaeobiogeography. *Palaeogeography, Palaeoclimatology, Palaeoecology*, **100**, 109–23.

Damborenea, S.E. and Manceñido, M.O. (1979). On the palaeogeographical distribution of the pectinid genus *Weyla* (Bivalvia, Lower Jurassic). *Palaeogeography, Palaeoclimatology, Palaeoecology*, **27**, 85–102.

Damborenea, S.E. and Mancẽido, M.O. (1992). A comparison of Jurassic marine benthonic faunas from South America and New Zealand. *Journal of the Royal Society of New Zealand*, **22**, 131–52.

Darlington, P.J. (1948). The geographical distribution of cold-blooded vertebrates. *Quarterly Review of Biology*, **23**, 1–26, 105–23.

Darlington, P.J. (1957). *Zoogeography : the geographic distribution of animals*. Wiley, New York.

Darwin, C.R. (1859). *The origin of species*. Murray, London.

Dercourt, J. *et al.* (1986), Geological evolution of the Tethys belt from the Atlantic to the Pamirs since the Lias. *Tectonophysics*, **123**, 241–315.

Dettman, M.E. (1989). Antarctica : Cretaceous cradle of austral temperate rainforests? In *Origins and evolution of the Antarctic biota* (ed. J.A. Crame), pp. 89–106, Geological Society, London.

Dewey, J.F. (1988). Lithospheric stress, deformation, and tectonic cycles: the disruption of Pangaea and the closure of Tethys. In *Gondwana and Tethys* (ed. M.G. Audley-Charles and A. Hallam), pp. 23–40. Oxford University Pres

Dhondt, A.V. (1992). Cretaceous inoceramid biogeography : a review. *Palaeogeography, Palaeoclimatology, Palaeoecology*, **92**, 217–32

Dilley, F.C. (1971). Cretaceous foraminiferal biogeography. In *Faunal provinces in space and time* (ed. F.A. Middlemiss, P.F. Rawson, and G. Newall), pp. 169–90. Seel House Press, Liverpool.

Dilley, F.C. (1973). Larger Foraminifera and seas through time. *Special Papers in Palaeontology*, **12**, 155–68.

Dimichele, W.A. and Aronson, R.B. (1992). The Pennsylvanian–Permian vegetational transition : a terrestrial analogue to the onshore–offshore hypothesis. *Evolution*, **46**, 807–24.

Donnelly, T.W. (1985). Mesozoic and Cenozoic plate evolution of the Caribbean region. In *The great American biotic interchange* (ed. F.G. Stehli and S.D. Webb), pp. 3–16. Plenum, New York.

Donovan, S.K.(ed.) (1989). *Mass extinction, processes and evidence.* Belhaven Press, London.

Doré, A.G. (1991). The structural foundation and evolution of Mesozoic seaways between Europe and the Arctic. *Palaeogeography, Palaeoclimatology, Palaeoecology*, **87**, 441–92.

Doyle, J.A. (1977). Patterns of evolution in early angiosperms. In *Patterns of evolution as illustrated by the fossil record* (ed. A. Hallam), pp. 501–46. Elsevier, Amsterdam.

Doyle, P. (1987). Lower Jurassic–Lower Cretaceous belemnite biogeography and the development of the Mesozoic Boreal Realm. *Palaeogeography, Palaeoclimatology, Palaeoecology*, **61**, 237–54.

Doyle, P. (1992). A review of the biogeography of Cretaceous belemnites. *Palaeogeography, Palaeoclimatology, Palaeoecology*, **92**, 207–16.

Duque-Caro, H. (1990). Neogene stratigraphy, paleoceanography and paleobiology in northwest South America and the evolution of the Panama Seaway. *Palaeogeography, Palaeoclimatology, Palaeoecology*, **77**, 203–34.

Durham, J.W. (1950). Cenozoic marine climates of the Pacific coast. *Bulletin of the Geological Society of America*, **61**, 1243–64.

Durham, J.W. and MacNeil, F.S. (1967). Cenozoic migrations of marine invertebrates through the Bering Strait region. In *The Bering land bridge* (ed. D.M. Hopkins), pp. 326–49. Stanford University Press, Stanford.

Edwards, D. (1990). Constraints on Silurian and Early Devonian phytogeographic analysis based on megafossils. In *Palaeozoic, palaeogeography and biogeography*, Geological Society Memoir no. 12. (ed. W.S. McKerrow and C.R. Scotese), pp. 233–42.

Ekman, S. (1953). *Zoogeography of the Sea.* Sidgwick and Jackson, London.

Eldredge, N. (1991). *The Miner's Canary.* Prentice Hall, New York.

Eldredge, N. and Ormiston, A.R. (1979). Biogeography of Silurian and Devonian trilobites of the Malvinokaffric Realm. In *Historical biogeography, plate tectonics, and the changing environment* (ed. J. Gray and A.J. Boucot), pp. 147–67. Oregon State University Press, Corvallis.

Enay, R. (1973). Upper Jurassic (Tithonian) ammonites. In *Altas of palaeobiogeography* (ed. A. Hallam), pp. 297–307. Elsevier, Amsterdam.

Estes, R. (1983). The fossil record and early distribution of lizards. In *Advances in herpetology and evolutionary biology* (ed. A. Rhodin and K. Miyata), pp. 365–98. Museum of Comparative Zoology, Cambridge, Massachusetts.

Estes, R. and Hutchison, J.H. (1980). Eocene lower vertebrates from Ellesmere Island, Canadian Arctic Archipelago. *Palaeogeography, Palaeoclimatology, Palaeoecology*, **30**, 325–47.

Fallaw, W.C. (1979*a*). A test of the Simpson Coefficient and other binary coefficients of faunal similarity. *Journal of Palaeontology*, **53**, 1029–34.

Fallaw, W.C. (1979*b*). Trans-North Atlantic similarity among Mesozoic and Cenozoic invertebrates correlated with widening of the ocean basin. *Geology*, **7**, 398–400.

Fallaw, W.C. (1983). Trans-Pacific faunal similarities among Mesozoic and Cenozoic invertebrates related to plate tectonic processes. *American Journal of Science*, **283**, 166–72.

Fallaw, W.C. and Dromgoole, E.L. (1980). Faunal similarities across the South Atlantic among Mesozoic and Cenozoic invertebrates correlated with widening of the ocean basin. *Journal of Geology*, **88**, 723–7.

Feldman, R.M. and Zinsmeister, W.J. (1984). New fossil crabs (Decapoda:

Brachyura) from the La Meseta Formation (Eocene) of Antarctica : palaeogeographic and biogeographic implications. *Journal of Paleontology*, **58**, 1046–61.

Finney, S.C. (1984). Biogeography of Ordovician graptolites in the southern Appalachians. In *Aspects of the Ordovician system* (ed. D.L. Bruton), pp. 167–76. Universitetsforlaget, Oslo.

Finney, S.C. (1986). Graptolite biofacies and correlation of eustatic, subsidence and tectonic events in the Middle to Upper Ordovician of North America. *Palaios* **1**, 435–61.

Finney, S.C. and Chen Xu (1990). The relationship of Ordovician graptolite provincialism to palaeogeography. In *Palaeozoic, palaeogeography and biogeography*, Geological Society Memoir no. 12.(ed. W.S, McKerrow and C.R. Scotese), pp. 123–8.

Fischer, A.G. (1960). Latitundinal variations in organic diversity. *Evolution*, **14**, 64–81.

Fischer, A.G. (1984). The two Phanerozoic supercycles. In *Catastrophes and earth history* (ed. W.A. Berggren and J.A. Van Couvering), pp. 129–50. Princeton University Press, Princeton.

Flessa, K.W. (1975). Area, continental drift and mammalian diversity. *Paleobiology*, **1**, 189–94.

Flessa, K.W. (1981). The regulation of mammalian faunal similarity among the continents. *Journal of Biogeography*, **8**, 427–37.

Flessa, K.W. and Hardy, M.C. (1988). Devonian conodont biogeography : quantitative analysis of provinciality. *Historical Biology*, **1**, 103–34.

Florin, R. (1963). The distribution of conifer and taxal genera in time and space. *Acta Horticultura Bergen*, **20**, 121–312.

Flynn, L.J., Tedfad, R.H., and Qiu Zhanxiang (1991). Enrichment and stability in the Pliocene mammalian fauna of North China. *Paleobiology*, **17**, 246–65.

Förster, R. (1978). Evidence for an open seaway between northern and southern proto- Atlantic in Albian times. *Nature*, **272**, 158–9.

Fortey, R.A. (1975). Early Ordovician trilobite communities. *Fossils and Strata*, **4**, 339–60.

Fortey, R.A. (1984). Global earlier Ordovician transgressions and regressions and their biological implications. In *Aspects of the Ordovician system* (ed. D.L. Bruton), pp. 37–50. Universitetsforlaget, Oslo.

Fortey, R.A. and Cocks, L.R.M. (1986). Marginal faunal belts and their structural implications, with examples from the Lower Palaeozoic. *Journal of the Geological Society*, **143**, 151–60.

Fortey, R.A. and Mellish, C.J.T. (1992). Are some fossils better than others for inferring palaeogeography? *Terra Nova*, **4**, 210–16.

Frakes, L.A., Francis, J.E., and Syktus, J.I. (1992). *Climate modes of the Phanerozoic*. Cambridge University Press, Cambridge.

Frey, R.W. and Seilacher, A. (1980). Uniformity in marine invertebrate ichnology. *Lethaia* **13**, 183–207.

Fujita, K. and Newberry, J.T. (1982). Tectonic evolution of northeastern Siberia and adjacent regions. *Tectonophysics*, **89**, 337–57.

Fürsich, F.T. and Sykes, R.M. (1977). Palaeobiogeography of the European Boreal Realm during Oxfordian (Upper Jurassic) times : a quantitative approach. *Neues Jahrbuch für Geologie und Paläontologie Abhandlungen*, **155**, 137–61.

Galton, P.M. (1977). The ornithopod dinosaur *Dryosaurus* and a Laurasia–Gondwanaland connection in the Upper Jurassic. *Nature*, **268**, 230–2.

Gasparini, Z., Fernandez, M., and Powell, J. (1993). New Tertiary sebecosuchians

(Crocodylomorpha) from South America : phylogenetic implications. *Historical Biology*, **7**, 1–19.

Gayet, M., Rage, J-C, Sempere, T., and Gagnier, P-Y (1992). Modalités des échanges de vertébrés continentaux entre l'Amerique du Nord et l'Amerique du Sud au Cretacé supérieur et au Paléocène. *Bulletin de la Société geologique de France* **163**, 781–91.

George, W. (1981). Wallace and his line. In *Wallace's line and plate tectonics* (ed. T.C. Whitmore), pp. 3–8. Clarendon Press, Oxford.

George, W. (1993). The strange rodents of Africa and South America. In *The Africa-South America connection* (ed. W. George and R. Lavocat), pp. 119–41. Clarendon Press, Oxford.

George, W. and Lavocat, R. (ed.) (1993). *The Africa–South America connection*. Clarendon Press, Oxford.

Gheerbrant, E. (1990). On the early biogeographical history of the African placentals. *Historical Biology*, **4**, 107–16.

Ghiold, J. and Hoffman, A. (1986). Biogeography and biogeographic history of clypeasteroid echinoids. *Journal of Biogeography*, **13**, 183–206.

Gingerich, P.D. (1985). South American mammals in the Paleocene of North America. In *The great American biotic interchange* (ed. F.G. Stehli and S.D. Webb), pp. 123–37. Plenum, New York.

Gingerich, P.D. (1986). Early Eocene *Cantius torresi* – oldest primate of modern aspect from North America. *Nature*, **319**, 319–21.

Gordon, W.A. (1970). Biogeography of Jurassic Foraminifera. *Bulletin of the Geological Society of America*, **81**, 1689–704.

Grande, L. (1985). The use of paleontology in systematics and biogeography and a time control refinement for historical biogeography. *Paleobiology*, **11**, 234–43.

Gray, J. and Boucot, A.J., ed. (1979). *Historical biogeography, plate tectonics, and the changing environment*. Oregon State University Press, Corvallis.

Haffer, J. (1969). Speciation in Amazonian forest birds. *Science*, **165**, 131–7.

Haffer, J. (1987). Quaternary history of tropical America. In *Biogeography and Quaternary history in tropical America* (ed. T.C. Whitmore and G.T. Prance), pp. 1–18. Clarendon Press, Oxford.

Hallam, A. (1967*a*). The bearing of certain palaeogeographic data on continental drift. *Palaeogeography, Palaeoclimatology, Palaeoecology*, **3**, 201–41.

Hallam, A. (1967*b*). Depth indicators in marine sedimentary environments. *Marine Geology*, **5**, 329–555.

Hallam, A. (1972). Diversity and density characteristics of Pliensbachian–Tarcian molluscan and brachiopod faunas of the North Atlantic margins. *Lethaia*, **5**, 389–412.

Hallam, A. (1973*a*). *A revolution in the earth sciences. Clarendon Press, Oxford.*

Hallam, A. (1973*b*). *Atlas of palaeobiogeography*. Elsevier, Amsterdam.

Hallam, A. (1974). Changing patterns of provinciality and diversity of fossil animals in relation to plate tectonics. *Journal of Biogeography*, **1**, 213–25.

Hallam, A. (1975). *Jurassic environments*. Cambridge University Press, Cambridge.

Hallam, A. (1977*a*). Jurassic bivalve biogeography. *Paleobiology*, **3**, 58–73.

Hallam, A. (1977*b*). Biogeographic evidence bearing on the creation of Atlantic seaways in the Jurassic. *Milwaukee Public Museum Special Publication, Biology and Geology*, No. 2, 23–34.

Hallam, A. (1981*a*). Relative importance of plate movements, eustasy and climate in controlling major biogeographic changes since the early Mesozoic. In *Vicariance biogeography : a Critique* (ed. G. Nelson and D.E. Rosen), pp. 303–40. Columbia University Press, New York.

Hallam, A. (1981*b*). *Facies interpretation and the stratigraphic record.* W.H. Freeman, Oxford.

Hallam, A. (1981*c*). Biogeographic relations between the northern and southern continents during the Mesozoic and Cenozoic. *Geologische Rundschau*, **70**, 583–95.

Hallam, A. (1983*a*). *Great geological controversies.* Oxford University Press.

Hallam, A. (1983*b*). Supposed Permo–Triassic megashear between Laurasia and Gondwana. *Nature*, **301**, 499–502.

Hallam, A. (1983*c*). Early and mid Jurassic molluscan biogeography and the establishment of the Central Atlantic Seaway. *Palaeogeography, Palaeoclimatology, Palaeoecology*, **43**, 181–93.

Hallam, A. (1984*a*). The unlikelihood of an expanding Earth. *Geological Magazine*, **131**, 653–5.

Hallam, A. (1984*b*). Distribution of fossil marine invertebrates in relation to climate. In *Fossils and climate* (ed. P.J. Brenchley), pp. 107–25. Wiley, Chichester.

Hallam, A. (1986). Evidence of displaced terranes from Permian to Jurassic faunas around the Pacific margins. *Journal of the Geological Society*, **143**, 209–16.

Hallam, A. (1990). Biotic and abiotic factors in the evolution of early Mesozoic marine molluscs. In *Causes of evolution : a paleontological perspective* (ed. R.M. Ross and W.D. Allmon), pp. 249–69. University of Chicago Press, Chicago.

Hallam, A. (1992). *Phanerozoic sea-level changes.* Columbia University Press, New York.

Hallam, A. (1993). Jurassic climates as inferred from the sedimentary and fossil record. *Philosophical Transactions of the Royal Society of London*, **B341**, 287–96.

Hallam, A., Perez, E., and Biro, L. (1986). Facies analysis of the Lo Valdes Formation (Tithonian–Hauterivian) of the High Cordillera of central Chile and the palaeogeographic evolution of the Andean Basin. *Geological Magazine*, **123**, 425–435.

Halle, T.G. (1937). The relation between the Late Palaeozoic floras of eastern and northern Asia. *Comptes Rendus de la Deuxième Congrès pour l'Avancement d'Étude Stratigraphique du Carbonifère, Heerlen, 1935*, pp. 237–45. Maastricht.

Hansen, T.A. (1987). Extinction of late Eocene to Oligocene molluscs : relationship to shelf area, temperature changes, and impact events. *Palaios*, **2**, 69–75.

Haq, B.U. (1980). Biogeographic history of Miocene calcareous nannoplankton and paleoceanography of the Atlantic Ocean. *Micropaleontology*, **26**, 414–43.

Haq, B.U., Premoli-Silva, I., and Lohman, G.P. (1977). Calcareous plankton paleobiogeographic evidence for major climatic fluctuations in the Early Cenozoic Atlantic Ocean. *Journal of Geophysical Research*, **82**, 3861–76.

Harland, W.B., Armstrong, R.L., Cox, A.V., Craig, L.E., Smith, A.G., and Smith, D.G. (1989). *A geological time scale 1989.* Cambridge University Press, Cambridge.

Harrison, C.G.A., Barron, E.J., and Hag, W.W. (1979). Mesozoic evolution of the Antarctic Penninsula and the southern Andes. *Geology*, **7**, 374–8.

Hayami, I. (1961). On the Jurassic pelecypod faunas of Japan. *Journal of the Faculty of Science, University of Tokyo*, section 2, 243–343.

Hayami, I. (1984). Jurassic marine bivalve faunas and biogeography in southeast Asia. *Geology and Palaeontology of Southeast Asia*, **25**, 229–37.

Heckel, P.H. (1972). Recognition of ancient shallow marine environments. *Special publication of the Society of Economic Paleontologists and Mineralogists*, no. 16, 226–86.

Heckel, P.H. and Witzke, B.J. (1979). Devonian world palaeogeography determined from distribution of carbonates and related lithic palaeoclimatic indicators. *Special Papers in Palaeontology* No. **23**, 99–123.

Henderson, R.A. and Heron, M.L. (1977). A probabilistic method of paleobiogeographic analysis. *Lethaia*, **10**, 1–15.

Hessler, R.R. and Sanders, H.L. (1967). Faunal diversity in the deep sea. *Deep-Sea Research*, **14**, 65–79.

Hillebrandt, A. von (1981). Kontinentalverschiebung und die paläozoogeographischen Beziehungen des südamerikanischen Lias. *Geologische Rundschau*, **70**, 570–82.

Hillebrandt, A. von, Westermann, G.E.G., Callomon, J.H., and Detterman, R.L. (1993). Ammonites of the circum-Pacific region. In *The Jurassic of the circum-Pacific* (ed. G.F.G. Westermann), pp. 342–59. Cambridge University Press, Cambridge.

Hoffstetter, R. (1972). Relationships, origins and history of the ceboid monkeys and caviomorph rodents : a modern reinterpretation. *Evolutionary Biology*, **6**, 323–47.

Hooker, J.D. (1853). *The botany of the antarctic voyage of H.M. Discovery ships Erebus and Terror* in the Years 1839–1843. II. *Flora Nova Zelandiae*. London.

Hooker, J.D. (1859). *The botany of the antarctic voyage of H.M. Discovery ships Erebus and Terror* in the Years 1839–1843. III. *Flora Tasmaniae*. London.

Hopkins, D.M. (1967). The Cenozoic history of Beringia – a synthesis. In *The Bering land bridge* (ed. D.M. Hopkins), pp. 451–84. Stanford University Press, Stanford.

Hottinger, L. (1973). Selected Paleogene larger foraminifera. In *Atlas of palaeobiogeography* (ed. A. Hallam), pp. 443–52. Elsevier, Amsterdam.

House, M.R. (1973). An analysis of Devonian goniatite distributions. *Special Papers in Palaeontology* no. 12, 305–17.

Howell, D.G. (ed.) (1985). *Tectonostratigraphic terranes of the circum-Pacific region*. Circum-Pacific Council for Energy and Mineral Resources, Houston. Earth Science Series no. 1.

Hubbard, R.N.L.B. and Boulter, M.C. (1983). Reconstruction of Palaeogene climate from palynological evidence. *Nature*, **301**, 147–50.

Huber, B.T. (1992). Paleobiogeography of Campanian–Maastrichtian foraminifera in the southern high latitudes. *Palaeogeography, Palaeoclimatology, Palaeoecology*, **92**, 325–60.

Hughes, C.P. (1973). Analysis of past faunal distributions. In *Implications of continental drift to the earth sciences*, (ed. D.H. Tarling and S.K. Runcorn), vol. 1, pp. 221–30. Academic Press, London.

Hughes, N.F. (ed.) (1973). *Organisms and continents through time. Special Papers in Palaeontology* no. 12.

Humphries, C.J. and Parenti, L.R. (1986). *Cladistic biogeography*. Clarendon Press, Oxford.

Humphries, C.J., Ladiges, P.Y., Roos, M., and Zandee, M. (1988). Cladistic biogeography. In *Analytical biogeography* (ed. A.A. Myers and P.S. Giller), pp. 371–404. Chapman and Hall, London.

Hutchinson, J.H. (1982). Turtle, crocodile, and champsosaur diversity changes in the Cenozoic of the north-central region of western United States. *Palaeogeography, Palaeoclimatology, Palaeoecology*, **37**, 149–64.

Imbrie, J. and Kipp, N.G. (1971). A new micropaleontological method for quantitative paleoclimatology : application to a late Pleistocene Caribbean core. In *The Late Cenozoic glacial ages* (ed. K.K. Turekian), pp. 71–179. Yale University Press, Newhaven.

Ishii, K., Okimura, Y., and Ichikawa, K. (1985). Notes on Tethys biogeography with reference to Middle Permian fusulinaceans. In *The Tethys* (ed. K. Nakazawa and J.M. Dickins), pp. 139–55. Tokai University Press, Tokyo.

Jablonski, D. (1986). Background and mass extinctions: the attenuation of macroevolutionary regimes. *Science*, **231**, 129–33.

Jablonski. D. (1989). The biology of mass extinctions: a palaeontological view. *Philosophical Transactions of the Royal Society*, **B325**, 357–68.

Jablonski, D. and Bottjer, D.J. (1990*a*). Onshore–offshore trends in marine invertebrate evolution. In *Causes of evolution : a paleontological perspective* (ed. R.M. Ross and W.D. Allmon), pp. 21–76. University of Chicago Press, Chicago.

Jablonski, D. and Bottjer, D.J. (1990*b*). The origin and diversification of major groups: environmental patterns and macroevolutionary lags. In *Major evolutionary radiations* (ed. P.D. Taylor and G.P. Larwood), pp. 17–57. Oxford University Press, Oxford.

Jablonski, D. and Lutz, R.A. (1983). Larval ecology of marine benthic invertebrates. *Biological Reviews*, **58**, 21–89.

Jablonski, D., Sepkoski, J.J., Bottjer, D.J., and Sheehan, P.M. (1983). Onshore – offshore patterns in the evolution of Phanerozoic shelf communities. *Science*, **222**, 1123–5.

Jackson, J.B.C., Jung, P., Coates, A.G., and Collins, L.S. (1993). Diversity and extinction of tropical American mollusks and emergence of the Isthmus of Panama. *Science*, **260**, 1624–6.

Jacobs, L.L., Cheema, I.U., and Shah, S.M.I. (1981). Zoogeographic implications of early Miocene rodents from the Bugti Beds, Baluchistan, Pakistan. *Geobios*, **15**, 101–3.

Jaeger, J.J., Coiffait, B., Tong, H., and Denys, C. (1987). Rodent extinctions following Messinian faunal exchanges between Western Europe and Northern Africa. *Mémoires de la Société Geologique de France*, **150**, 153–58.

Jaeger, J.J., Courtillot, V., and Tapponnier, P. (1989). Paleontological view of the ages of the Deccan Traps, and the Cretaceous/Tertiary boundary, and the India–Asia collision. *Geology*, **17**, 316–19.

Janis, C.M. (1989). A climatic explanation for patterns of evolutionary diversity in ungulate mammals. *Palaeontology*, **32**, 463–82.

Janis, C.M. and Damuth, J. (1990). Mammals. In *Evolutionary trends* (ed. K.J. McNamara), pp. 301–45. University of Arizona Press, Tucson.

Jansa, C.F. (1991). Processes affecting paleogeography, with examples from the Tethys. *Palaeogeography, Palaeoclimatology, Palaeoecology*, **87**, 345–71.

Jell, P.A. (1974). Faunal provinces and possible planetary reconstruction of the Middle Cambrian. *Journal of Geology*, **82**, 319–50.

Johnson, J.G. and Boucot, A.J. (1973). Devonian brachiopods. In *Atlas of palaeobiogeography* (ed. A. Hallam), pp. 89–96. Elsevier, Amsterdam.

Johnson, J.G., Klapper, G., and Sandberg, C.A. (1985). Devonian eustatic fluctuations in Euramerica. *Bulletin of the Geological Society of America*, **96**, 567–87.

Jones, D.S. and Hasson, P.F. (1985). In *The great American biotic interchange* (ed. F.G. Stehli and S.D. Webb), pp. 325–55. Plenum Press, New York

Kaiho, K. (1992). Comparative taxonomy and faunal provinces of benthic foraminifera from late Eocene intermediate water. *Micropaleontology*, **38**, 363–96.

Kaljo, D. and Klaamann, E. (1973). Ordovician and Silurian corals. In *Atlas of palaeobiogeography* (ed. A. Hallam), pp. 37–45. Elsevier, Amsterdam.

Kauffman, E.G. (1973). Cretaceous Bivalvia. In *Atlas of palaeobiogeography* (ed. A. Hallam), pp. 353–83. Elsevier, Amsterdam.

Kauffman, E.G. and Scott, R.W. (1976). Basic concepts of community ecology and paleoecology. In *Structure and classification of paleocommunities* (ed. R.W. Scott and R.R. West), pp. 1–28. Dowden, Hutchinson and Ross, Stroudsburg, Pennsylvania.

Kearey, P. and Vine, F.J. (1990). *Global tectonics*. Blackwell Scientific Publications, Oxford.

Keast, A. (1973). Contemporary biotas and the separation sequence of the southern continents. In *Implications of continental drift to the earth sciences* (ed. D.H. Tarling and S.K. Runcorn), pp. 309–43. Academic Press, London.

Keigwin, L.D. (1978). Pliocene closing of the isthmus of Panama, based on biostratigraphic evidence from nearby Pacific Ocean and Caribbean sea cores. *Geology*, **6**, 630–4.

Keller, G. (1983). Paleoclimatic analysis of middle Eocene through Oligocene planktonic foraminiferal faunas. *Palaeogeography, Palaeoclimatology, Palaeoecology*, **43**, 73–94.

Kelley, P.H., Raymond, A., and Lutken, C.B. (1990). Carboniferous brachiopod migration and latitudinal diversity: a new palaeoclimatic model. In *Palaeozoic, palaeogeography and biogeography, Geological Society Memoir no. 12.* (ed. W.S. McKerrow and C.R. Scotese), pp. 325–332.

Kemp, E.M. (1978). Tertiary climatic evolution and vegetation history in the southeast Indian Ocean region. *Palaeogeography, Palaeoclimatology, Palaeoecology*, **24**, 169–208.

Kennedy, W.J. and Cooper, M.R. (1975). Cretaceous ammonite distributions and the opening of the South Atlantic. *Journal of the Geological Society*, **131**, 283–8.

Kennett, J.P. (1978). The development of planktonic biogeography in the Southern Ocean during the Cenozoic. *Marine Micropaleontology*, **3**, 301–45.

Kennett, J.P. (1980). Paleoceanographic and biogeographic evolution of the Southern Ocean during the Cenozoic, and Cenozoic microfossil datuns. *Palaeogeography, Palaeoclimatology, Palaeoecology*, **31**, 123–52.

Kennett, J.P. and Barker, P.F. (1990). Latest Cretaceous to Cenozoic climate and oceanographic developments in the Weddell Sea, Antarctica : an ocean drilling perspective. *Proceedings of the Oceanic Drilling Project*, **113**, 937–60.

Kennett, J.P. and Stott, L.D. (1991). Abrupt deep-sea warming, palaeoceanographic changes and benthic extinctions at the end of the Palaeocene. *Nature*, **353**, 225–9.

Kennett, J.P., Keller, G., and Srinivasan, M.S. (1985). Miocene planktonic foraminiferal biogeography and paleoceanographic development of the Indo-Pacific region. *Geological Society of American Memoir*, **163**, 197–236.

Kielan–Jaworowska, Z. (1974). Migrations of the Multituberculata and the Late Cretaceous connections between Asia and North America. *Annals of the South African Museum*, **64**, 231–43.

Kimura, T. (1984). Mesozoic floras of East and Southeast Asia, with a short note on the Cenozoic floras of Southeast Asia and China. *Geology and Palaeontology of Southeast Asia*, **25**, 325–50.

King, G.M. (1991). Terrestrial tetrapods and the end-Permian event: a comparison of analyses. *Historical Biology*, **5**, 239–55.

Klapper, G. and Johnson, J.G. (1980). Endemism and dispersal of Devonian conodonts. *Journal of Paleontology*, **54**, 400–55.

Klitgard, K.D. and Schouten, H. (1986). Plate kinematics of the central Atlantic. In *The geology of North America, M* (ed. P.R. Vogt and B.E. Tucholke), pp. 351–78. Geological Society of America, Boulder, Colorado.

Knoll, A.H. (1984). Patterns of extinction in the fossil record of vascular plants. In *Extinctions* (ed. M. Nitecki), pp. 21–68. University of Chicago Press, Chicago.

Kobayashi, T. and Tamura, M. (1984). The Triassic Bivalvia of Malaysia, Thailand and adjacent areas. *Geology and Palaeontology of Southeast Asia* **25**, 201–27.

Koch, C.F. (1987). Prediction of sample size effects on the measured temporal and geographic distribution patterns of species. *Paleobiology*, **13**, 100–7.

Koutsoukos, E.A.M. (1992). Late Aptian to Maastrichtian foraminiferal biogeography and palaeoceanography of the Sergipe Basin, Brazil. *Palaeogeography, Palaeoclimatology, Palaeoecology*, **92**, 295–324.

Kristan-Tollmann, E. (1988). Unexpected microfaunal communities within the Triassic Tethys. In *Gondwana and Tethys* (ed. M.G. Audley-Charles and A. Hallam), pp. 213– 23. Oxford University Press, Oxford.

Krömmelbein, K. (1965a). Neue, für Vergleiche mit Westafrika wichtige, Ostracoden-Arten der brasilianischen Bahia-Serie (Ober Jura/Unter Kreide in Wealden Fazies). *Senckenbergiana Lethaia*, **46**, 1772–213.

Krömmelbein, K. (1965b). Ostracoden aus der nicht-marinen Unter Kreide (West–Afrikanischer Wealden) des Congo-Kustenbeckens. *Meyniana*, **15**, 59–74.

Krömmelbein, K. (1971). Non-marine Cretaceous ostracodes and their importance for the hypothesis of Gondwanaland. *Proceedings of the 2nd IUGS Gondwana Symposium*, pp. 617–19. Pretoria.

Kummel, B. (1973). Lower Triassic (Scythian) molluscs. In *Atlas of palaeobiogeography* (ed. A. Hallam), pp. 225–33. Elsevier, Amsterdam.

Kurtén, B. (1966). Holarctic land connections in early Tertiary. *Commentationes Biologia Societa Scientifica Fennica*, **29**, 1–5.

Kurtén, B. (1973). Some Early Tertiary land mammals. In *Atlas of palaeobiogeography* (ed. A. Hallam), pp. 437–42. Elsevier, Amsterdam.

Larsen, R.L. (1991). Geological consequences of superplumes. *Geology*, **19**, 963–6.

Larwood, G.P. ed. (1988). *Extinction and survival in the fossil record*. Clarendon Press, Oxford.

Laufeld, S. (1979). Biogeography of Ordovician Silurian and Devonian Chitinozoa. In *Historical biogeography, plate tectonics, and the changing environment* (ed. J. Gray and A.J. Boucot), pp. 75–90. Oregon State University Press, Corvallis.

Laveine, J.P., Lemoigne, Y., Li, X., Wu, X., Zhang, S., Zhao, X., Zha,

W., and Zha, J. (1987). Paléogéographie de la Chinè au Carbonifère à la lumière des données paléobotaniques, par comparison avec les assemblages carbonifères d'Europe occidentale. *Comptes Rendus de l'Academie des Sciences, de Paris*, **304**, Sér. II, 391–94.

Lavocat, R. (1974). The interrelationships between the African and South American rodents and their bearing on the problem of the origin of the South American monkeys. *Journal of Human Evolution*, **3**, 523–6.

Levinton, J.S. (1988). *Genetics, palaontology, and evolution*. Cambridge University Press, Cambridge.

Li, Xiaochi and Grant-Mackie, J.A. (1993). Jurassic sedimentary cycles and eustatic sea-level changes in southern Tibet. *palaeogeography, Palaeoclimatology, Palaeoecology*, **101**, 27–48.

Liao, W.H. (1990). The biogeographic affinities of East Asian corals. *In Palaeozoic, Palaeogeography and Biogeography*, Geological Society Memoir no. 12. (ed. W.S. McKerrow and C.R. Scotese), pp. 175– 9.

Lin, J.L. and Watts, D.R. (1988). Palaeomagnetic constraints on Himalyayan–Tibetan tectonic evolution. *Philosophical Transactions of the Royal Society*, A **326**, 177–88.

Lin, J.L., Fuller, M., and Zhang, W.Y. (1985). Preliminary Phanerozoic polar wander paths for the North and South China blocks. *Nature*, **313**, 444–9.

Lindberg, D.R. (1991). Marine biotic interchange between the northern and southern hemispheres. *Paleobiology*, **17**, 308–24.

Lindsay, E.H., Opdyke, N.D., and Johnson, N.M. (1980). Pliocene dispersal of the horse *Equus* and late Cenozoic mammalian dispersal events. *Nature*, **287**, 135–8.

Lindsay, E.H., Opdyke, N.D., and Johnson, N.M. (1984). Blancan–Hemphillian land mammal ages and late Cenozoic mammal dispersal events. *Annual Reviews of Earth and Planetary Sciences*, **12**, 445–88.

Lomolino, M.V. (1985). Body size of mammals on islands: the island rule re-examined. *American Naturalist*, **125**, 310–16.

Lynch, J.D. (1988). Refugia. In *Analytical biogeography* (ed. A.A. Myers and P.S. Giller), pp. 311–42. Chapman and Hall, London.

McArthur, R.H. and Wilson, E.O. (1967). *The theory of island biogeography*. Princeton University Press, Princeton.

McElhinny, M.W., Taylor, S.R., and Stevenson, D.J. (1978). Limits to the expansion of Earth, Moon, Mars and Mercury and to changes in the gravitational constant. *Nature*, **271**, 316–21.

McKenna, M.C. (1973). Sweepstakes, filters, corridors, Noah's Arks, and beached Viking Funeral Ships in palaeogeography. In *Implications of continental drift to the earth sciences* (ed. D.H. Tarling and S.K. Runcorn), pp. 293–308. Academic Press, London.

McKenna, M.C. (1975). Fossil mammals and Early Eocene North Atlantic land continuity. *Annals of the Missouri Botanic Garden*, **62**, 335–53.

McKenna, M.C. (1980). Eocene paleolatitude, climate, and mammals of Ellesmere Island. *Palaeogeography, Palaeoclimatology, Palaeoecology*, **30**, 349–62.

McKenna, M.C. (1983). Holarctic landmass rearrangements, cosmic events, and Cenozoic terrestrial organisms. *Annals of the Missouri Botanic Garden*, **70**, 459–89.

McKerrow, W.S. and Cocks, L.R.M. (1976). Progressive faunal migration across the Iapetus Ocean. *Nature*, **263**, 304–6.

McKerrow, W.S. and Cocks, L.R.M. (1986). Oceans, island arcs and olistostromes: the use of fossils in distinguishing sutures, terranes and environments around the Iapetus Ocean. *Journal of the Geological Society*, **143**, 185–91.

McKerrow, W.S. and Scotese, C.R. (ed). (1990). *Palaeozoic Palaeogeography and biogeography*. Geological Society Memoir No. 12.

McKinnon, T.C. (1983). Origin of the Torlesse terrane and coeval rocks, South Island, New Zealand. *Bulletin of the Geological Society of America*, **94**, 967–85.

Malmgren, B.A. and Haq, B.U. (1982). Assessment of quantitative techniques in paleobiogeography. *Marine Micropaleontology*, **7**, 213–36.

Manceñido, M.O. (1993). First record of Jurassic nucleatid brachiopods from the southwest Pacific with comments on the global distribution of the group. *Palaeogeography, Palaeoclimatology, Palaeoecology*, **100**, 189–207.

Marshall, L.G. (1985). Geochronology and land-mammal biochronology of the Transamerican faunal interchange. In *The great American biotic interchange* (ed. F.G. Stehli and S.D. Webb), pp 49–85. Plenum Press, New York.

Marshall, L.G., Webb, S.D., Sepkoski, J.J., and Raup, D.M. (1982). Mammalian evolution and the Great American Interchange. *Science*, **215**, 1351–7.

Martin, P.S. (1984). 'Prehistoric overkill: the global model'. In *Quaternary extinctions: a prehistoric revolution* (ed. P.S. Martin and R.G. Klein), pp. 354–403. University of Arizona Press, Tucson.

Matsuda, T. (1985). Late Permian to Early Triassic conodont, paleobiogeography in the 'Tethys Realm'. In *The Tethys* (ed. K. Nakazawa and J.M. Dickins), pp. 157–70. Tokai University Press, Tokyo.

Matsumoto, T. (1973). Late Cretaceous Ammonoidea. In *Atlas of palaeobiogeography* (ed. A. Hallam), pp. 421–9. Elsevier, Amsterdam.

Mattauer, J., Matte, P., Malavieille, J., Tapponnier, P., Maluski, H., Xu, Z., *et al*. (1985). *Nature*, **317**, 496–500.

Matthew, W.D. (1915). Climate and evolution. *Annals of the New York Academy of Sciences*, **24**, 171–218.

Maxwell, W.D. (1992). Permian and Early Triassic extinction of non-marine tetrapods. *palaeontology*, **35**, 571–84.

Metcalfe, I. (1988). Origin and assembly of south-east Asian continental terranes. In *Gondwana and Tethys* (ed. M.G. Audley-Charles and A. Hallam), pp. 101–18. Oxford University Press, Oxford.

Metcalfe, I. (1991). Late Palaeozoic and Mesozoic palaeogeography of Southeast Asia. *Palaeogeography, Palaeoclimatology, Palaeoecology*, **87**, 211–21.

Middlemiss, F.A., Rawson, P.F., and Newall, G. (1971). *Faunal provinces in space and time*. Seel House Press, Liverpool.

Milner, A.R. and Panchen, A.L. (1973). Geographical variation in the tetrapod faunas of the Upper Carboniferous and Lower Permian. In *Implications of continental drift to the earth sciences* (ed. D.H. Tarling and S.K. Runcorn), vol. 1, pp. 353–68. Academic Press, London.

Mizutani, S. and Kojima, S. (1992). Mesozoic radiolarian biostratigraphy of Japan and collage tectonics along the eastern continental margin of Asia. *Palaeogeography, Palaeoclimatology, Palaeoecology*, **96**, 3–22.

Molnar, P. and Tapponnier, P. (1975). Cenozoic tectonics of Asia: effects of a continental collision. *Science*, **189**, 419–26.

Monger, J.W.H. and Ross, C.A. (1971). Distribution of fusulinaceans in the western Canadian Cordillera. *Canadian Journal of Earth Sciences*, **8**, 259–78.

Morel, P. and Irving, E. (1981). Palaeomagnetism and the evolution of Pangaea. *Journal of Geophysical Research*, **86**, 1858–72.

Mutterlose, J. (1986). Upper Jurassic belemnites from the Orville Coast, Western Antarctica, and their palaeobiogeographic significance. *British Antarctic Survey Bulletin*, **70**, 1–22.

Mutterlose, J. (1992*a*). Biostratigraphy and palaeobiogeography of Early Cretareous calcareous nannofossils. *Cretaceous Research*, **13**, 167–89.

Mutterlose, J. (1992*b*). Migration and evolution patterns of floras and faunas in marine Early Cretaceous sediments of N.W. Europe. *Palaeogeography, Palaeoclimatology, Palaeoecology*, **94**, 261–82.

Myers, A.A. and Giller, P.S. (1988*a*). Preface. In *Analytical biogeography* (ed. A.A. Myers and P.S. Giller), pp xi–xiii. Chapman and Hall, London.

Myers, A.A. and Giller, P.S. (1988*b*). Process pattern and scale in biogeography. In *Analytical biogeography* (ed. A.A. Myers and P.S. Giller), pp. 3–12. Chapman and Hall, London.

Myers, G.S. (1938). Fresh-water fishes and West Indian zoogeography. *Annual Report of the Smithsonian Institution*, 1937, 339–44.

Myers, G.S. (1953*a*). Ability of amphibians to cross sea barriers, with especial reference to Pacific zoogeography. *Proceedings of the 7th Pacific Science Congress, New Zealand*, **4**, 19–27.

Myers, G.S. (1953*b*). Palaeogeographical significance of fresh-water fish distribution in the Pacific. *Proceedings of the 7th Pacific Science Congress, New Zealand*, **4**, 38–48.

Nakamura, K., Shimizu, D., and Zhuo-Ting, L. (1985). Permian palaeobiogeography of brachiopods based on the faunal provinces. In *The Tethys* (ed). K. Nakazawa and J.M. Dickins), pp. 185–98. Tokai University Press, Tokyo.

Nelson, G. (1978). From Candolle to Croizat: comments on the history of biogeography. *Journal of the History of Biology*, **11**, 269–305.

Nelson, G. and Platnick, N. (1981). *Systematics and biogeography: cladistics and vicariance*. Columbia University Press, New York.

Nelson, G. and Rosen, D.E. (ed.) (1981). *Vicariance biogeography: a critique*. Columbia University Press, New York.

Nestor, H. (1990). Biogeography of Silurian stromatoporoids. In *Palaeozoic, palaeogeography and biogeography*, Geological Society Memoir no. 12 (ed. W.S. McKerrow and C.R. Scotese), pp. 215– 221.

Neumayr, M. (1883). Über klimatische Zonen während der Jura-und Kreidzeit. *Königlische Akademie der Wissenschaft Wien Denkschrift*, **47**, 277–310.

Newton, C.R. (1987). Biogeographic complexity in Triassic bivalves of the Wallowa terrane, northwestern United States: oceanic islands, not continents, provide the best analogues. *Geology*, **15**, 1126–29.

Newton, C.R. (1988). Significance of Tethyan fossils in the American Cordillera. *Science*, **242**, 385–91.

Nie, S., Rowley, D.B., and Ziegler, A.M. (1990). Constraints on the location of Asian microcontinents in Palaeo-Tethys during the Late Palaeozoic. In *Palaeozoic, palaepogeography and biogeography*, Geological Society Memoir no. 12. (ed. W.S. McKerrow and C.R Scotese), see pp. 397–409.

Niklas, K.J., Tiffney, B.H., and Knoll, A.H. (1985). Patterns of vascular plant diversification : an analysis at the species level. In *Phanerozoic diversity patterns* (ed. J.W. Valentine), pp. 97–128. Princeton University Press, Princeton.

Nitecki, M.H. (ed.) (1984). *Extinctions*. University of Chicago Press, Chicago.

Norton, I.O. and Sclater, J.G. (1979). A model for the evolution for the Indian Ocean and the breakup of Gondwanaland. *Journal of Geophysical Research*, **84**, 6803–30.

Nur, A. and Ben-Avraham, Z. (1977). Lost Pacifica continent. *Nature*, 270, 41–3.

Nur, A. and Ben-Avraham, Z. (1981). Lost Pacifica continent: a mobilistic interpretation. In *Vicairance biogeography : a critique* (ed. G. Nelson and D.E. Rosen), pp. 3451–358. Columbia University Press, New York.

Okimura, Y., Ishii, K., and Ross, C.A. (1985). Biostratigraphic significance and faunal provinces of Tethyan Late Permian smaller Foraminifera. In *The Tethys* (ed. K. Nakazawa and J.M. Dickins), pp. 15–55. Tokai University Press, Tokyo.

Oliver, W.A. (1976). Biogeography of Devonian rugose corals. *Journal of Paleontology*, **50**, 365–73.

Oliver, W.A. and Pedder, A.E.H. (1979). Biogeography of Late Silurian and Devonian rugose corals in North America. In *Historical biogeography, plate tectonics and the changing environment* (ed. J. Gray and A.J. Boucot), pp. 131–45. Oregon State University Press, Corvallis.

Olsson, R.K. (1977). Mesozoic foraminifera-western Atlantic. In *Stratigraphic micropaleontology of Atlantic basin and borderlands* (ed. F.M. Swain), pp. 205–21. Elsevier, Amsterdam.

Owen, H.G. (1976). Continental displacement and expansion of the Earth during the Mesozoic and Cen ıc. *Philosophical Transactions of the Royal Society of London*, A **281**, 223–91.

Padian, K. and Clemens, W.A. (1985). Terrestrial vertebrate diversity: episodes and insights. In *Phanerozoic diversity patterns* (ed. J.W. Valentine), pp. 41–96. Princeton University Press, Princeton.

Palmer, A.R. (1973). Cambrian trilobites. In *Atlas of palaeobiogeography* (ed. A. Hallam), pp. 3–11. Elsevier, Amsterdam.

Paris, F. and Robardet, M. (1992). Early Palaeozoic palaeobiogeography of the Variscan regions. *Tectonophysics*, **177**, 193–213.

Parrish, J.M., Parrish J.T., Hutchinson, J.H., and Spicer, R.A. (1987). Late Cretaceous vertebrate fossils from the North Slope of Alaska and implications for dinosaur ecology. *Palaios*, **2**, 377–89.

Parrish, J.T. (1993). The palaeogeography of the opening South Atlantic. In *The Africa–South America connection* (ed. W. George and R. Lavocat), pp. 8–27. Clarendon Press, Oxford.

Patterson, C. (1983). Aims and methods in biogeography. In *Evolution, time and space : the emergence of the biosphere* (ed. R.W. Sims, J.H. Price, and P.E.S. Whalley), pp. 1–28, Systematics Association Special Volume 23.

Patterson, B. and Pascual, R. (1972). The fossil mammal fauna of South America. In *Evolution, mammals and southern continents* (ed. A. Keast, F.C. Erk, and B. Glass), pp. 247–309. State University of New York Press, Albany.

Pedder, A.E.H. and Oliver, W.A. (1990). Rugose coral distribution as a test of Devonian palaeogeographic models. In *Palaeozoic, palaeogeography and biogeography* Geological Society memoir no. 12. (ed). W.S. McKerrow and C.R. Scotese), pp. 267–75.

Pickering, K.T., Bassett, M.G., and Siveter, D.J. (1988). Late Ordovician–early Silurian destruction of the Iapetus Ocean: Newfoundland, British Isles and Scandinavia : a discussion. *Transactions of the Royal Society of Edinburgh*, **79**, 361–82.

Pindell, J.L., Cande, S.C., Pitman, W.C., Rowley, D.B., Dewey, J.F., LaBrecque, J., and Haxby, W. (1988). A plate-kinematic framework for models of Caribbean evolution. *Tectonophysics*, **155**, 121–38.

Platnick, N. and Nelson, G. (1978). A method of analysis for historical biogeography. *Systematic Zoology*, **27**, 1–16.

Plumstead, E.P. (1973). The Late Palaeozoic Glossopteris Flora. In *Atlas of palaeobiogeography* (ed. A. Hallam), pp. 187–205. Elsevier, Amsterdam.

Powell, C., McA and Conaghan, P.J. (1973). Plate tectonics and the Himalayas. *Earth and Planetary Science Letters*, **20**, 1–12.

Prasad, G.V.R. and Rage, J.C. (1991). A discoglossid frog in the latest Cretaceous (Maastrichtian) of India. Further evidence for a terrestrial route between India and Laurasia in the latest Cretaceous. *Comptes Rendus de l'Academie des Sciences de Paris*, **304** sér. II, 391–4.

Prasad, G.V.R. and Sahni, A. (1988). First Cretaceous mammal from India. *Nature*, **332**, 638–40.

Prothero, D.R. (1989). Stepwise extinctions and climatic decline during the later Eocene and Oligocene. In *Mass extinctions: processes and evidence* (ed. S.K. Donovan), pp. 217–34. Belhaven Press, London.

Prothero, D.R. and Estes, R. (1980). Late Jurassic lizards from Corno Bluff, Wyoming and their palaeobiogeographic significance. *Nature*, **286**, 484–6.

Rage, J.C. (1988). Gondwana, Tethys, and terrestrial vertebrates during the Mesozoic and Cainozoic. In *Gondwana and Tethys* (ed. M.G. Audley-Charles and A. Hallam), pp. 255–73. Oxford University Press, Oxford.

Raup, D.M. (1976). Species diversity in the Phanerozoic. *Paleobiology*, **2**, 279–88.

Raup, D.M. (1991). *Extinction: bad genes or bad luck?* Norton, New York.

Raup, D.M. and Crick, R.E. (1979). Measurement of faunal similarity in paleontology. *Journal of Paleontology*, **53**, 1213–27.

Raup, D.M. and Sepkoski, J.J. (1982). Mass extinctions in the marine fossil record. *Science*, **215**, 1501–3.

Raven, P.H. (1979). Plate tectonics and southern hemisphere biogeography. In *Tropical botany* (ed. K. Larsen and L.B. Holm-Nielsen), pp. 3–23. Academic Press, London.

Raven, P.H. and Axelrod, D.I. (1974). Angiosperm biogeography and past continental movements. *Annals of the Missouri Botanic Garden*, **61**, 539–673.

Raymond, A. (1985). Floral diversity, phytogeography, and climatic amelioration during the Early Carboniferous. *Paleobiology*, **11**, 293–309.

Raymond, A. (1987). Paleogeographic distribution of Early Devonian plant traits. *Palaios*, **2**, 113–32.

Reid, R.E.H. (1973). Origin of the Mesozoic 'Boreal' realm. *Geological Magazine*, **110**, 67–9.

Repenning, C.A. (1967). Palearctic–Nearctic mammalian dispersal in the Late Cenozoic. In *The Bering land bridge* (ed. D.M. Hopkins), pp. 451–84. Stanford University Press, Stanford.

Riccardi, A.C. (1991). Jurassic and Cretaceous marine connections between the Southeast Pacific and Tethys. *Palaeogeography, Palaeoclimatology, Palaeoecology*, **87**, 155–89.

Rich, P.V., Rich, T.H., Wagstaff, B.E., Mason, J.M., Douthill, C.B., Gregory, R.T., and Felton, E.A. (1988). Evidence for low temperatures and biological diversity in Cretaceous high latitudes of Australia. *Science*, **242**, 1403–6.

Richter, R. and Richter, E. (1942). Die Trilobiten der Weismes-schichten am Hohen

Venn, mit Bermerkungen über die Malvinocaffrische Provinz. *Senckenbergische Naturforschung Gesellschaft Abhandlungen*, **25**, 156–79.

Rögl, F. and Steiniger, F.F. (1984). Neogene Paratethys, Mediterranean and Indo-Pacific seaways: implications for the paleobiogeography of marine and terrestrial biotas. In *Fossils and climate* (ed. P.J. Brenchley), pp. 171–200. Wiley, Chichester.

Romine, K. (1985). Radiolarian biogeography and paleoceanography of the North Pacific at 8 Ma. *Geological Society of America Memoir*, **163**, 237–73.

Rosen, B.R. (1984). Reef coral biogeography and climate through the Late Cainozoic: just islands in the sun or a critical pattern of islands? In *Fossils and climate* (ed. P.J. Brenchley), pp. 201–62. Wiley, London.

Rosen, B.R. (1988). From fossils to earth history : applied historical biogeography. In *Analytical biogeography* (ed. A.A. Myers and P.S. Giller), pp. 437–81. Chapman and Hall, London.

Rosen, B.R. and Smith, A.B. (1988). Tectonics from fossils? Analysis of reef-coral and sea-urchin distributions from late Cretaceous to Recent, using a new method. In *Gondwana and Tethys* (ed. M.G. Audley-Charles and A. Hallam), pp. 275–306. Oxford University Press, Oxford.

Rosenzweig, M.L. and McCord, R.D. (1991). Incumbent replacement: evidence for long-term evolutionary progress. *Paleobiology*, **17**, 202–13.

Ross, C.A. and Ross, J.R.P. (1985). Carboniferous and Early Permian biogeography. *Geology*, **13**, 27–30.

Ross, J.R.P. (1978). Biogeography of Permian ectoproct Bryozoa. *Palaeontology*, **21**, 346– 56.

Ross, J.R.P. and Ross, C.A. (1990). Late Palaeozoic bryozoan biogeography. In *Palaeozoic, palaeogeography and biogeography*, Geological Society Memoir no. 12. (ed. W.S. McKerrow and C.R. Scotese), pp. 353– 62.

Ross, M.I. and Scotese, C.R. (1988). A hierarchical tectonic model of the Gulf of Mexico and Caribbean region. *Tectonophysics* **155**, 139–68.

Roth, P.H. and Krumbach, K.R. (1986). Middle Cretaceous calcareous nanno-fossil biogeography and preservation in the Atlantic and Indian Oceans: implications for paleoceanography. *Marine Micropaleontology*, **10**, 235–66.

Roth, V.L. (1992). Inferences from allometry and fossils: dwarfing of elephants on islands. *Oxford Surveys in Evolutionary Biology*, **8**, 259–88.

Rowley, D.B. and Lottes, A.L. (1988). Plate-kinematic reconstructions of the North Atlantic and Arctic: Late Jurassic to present. *Tectonophysics*, **155**, 73–120.

Ruddiman, W.F. and Kutzbach, J.E. (1989). Forcing of late Cenozoic northern Hemisphere climate by plateau uplift in southeast Asia and the American southwest. *Journal of Geophysical Research*, **94**, 18409–27.

Ruddiman, W.F. and McIntyre, A. (1984). Ice-age thermal response and climatic role of the surface Atlantic Ocean, 40°N to 63°N. *Bulletin of the Geological Society of America*, **95**, 381–96.

Ruddiman, W.F., Prell, W.L., and Raymo, M.E. (1989). History of the late Cenozoic uplift on south-east Asia and the American southwest: rationale for general circulation modelling experiments. *Journal of Geophysical Research*, **94**, 18379–91.

Sahni, A. (1984). Cretaceous–Paleocene terrestrial faunas of India: lack of endemism during drifting of the Indian Plate. *Science*, **226**, 441–3.

Sahni, A. and Kumar, K. (1974). Palaeogene palaeobiogeography of the Indian subcontinent. *Palaeogeography, Palaeoclimatology, Palaeoecology* **15**, 209–26.

Sahni, A., Bhatia, S.B., Hartenberger, J.L., Jaeger, J.J., Kumar, K., Sudre, J., and Vianey-Liaud, M. (1981). Vertebrates from the Sabatha formation and comments on the biogeography of Indian subcontinent during the early Paleogene. *Bulletin de la Société géologique de France*, **23**, 689–95.

Sahni, A., Kumar, K., Hartenberger, J.L., Jaeger, J.J., Rage, J.C., Sudre, J., and Vianey-Liaud, M. (1982). Microvertébrés nouveaux des Trapps du Deccan (Inde): mise en évidence d'une voie de communication terrestre probable entre la Laurasie et de l'Inde à la limite Cretacé–Tertiaire. *Bulletin de la Société géologique de France*, **24**, 1093–9.

Sanders, H.L. (1968). Marine benthic diversity: a comparative study. *American Naturalist*, **102**, 243–82.

Sandy, M.R. (1991). Aspect of Middle–Late Jurassic–Cretaceous Tethyan brachiopod biogeography in relation to tectonic and palaeoceanographic developments. *Palaeogeography, Palaeoclimatology, Palaeoecology*, **87**, 137–54.

Sato, T. (1962). Etudes biostratigraphiques des ammonites du Jurassique du Japon. *Société géologique de France Memoire*, **94**.

Savostin, L.A., Sibuet, J.C., Zonenshain, L.P., Le Pichon, X., and Roulet, M.-J. (1986). Kinematic evolution of the Tethys belt from the Atlantic Ocean to the Pamirs since the Triassic. *Tectonophysics*, **123**, 1–35.

Schallreuter, R.E.L. and Siveter, D.J. (1985). Ostracodes across the Iapetus Ocean. *Palaeontology*, **28**, 577–98.

Scheltema, R.S. (1971). The dispersal of the larvae of shoal-water benthic invertebrate species over long distances by ocean currents. In *Fourth European Marine Biology Symposium* (ed. D.J. Crisp), pp. 7–28. Cambridge University Press, Cambridge.

Scheltema, R.S. (1977). Dispersal of marine invertebrate organisms: paleobiogeographic and biostratigraphic implications. In *Concepts and methods of biostratigraphy* (ed. E.G. Kauffman and J.E. Hazel), pp. 73–108. Dowden, Hutchinson and Ross, Stroudsburg, Pennsylvania.

Schermer, E.R., Howell, D.G., and Jones, D.L. (1984). The origin of allochthonous terranes: perspectives on the growth and shaping of continents. *Annual Reviews of Earth and Planetary Sciences*, **12**, 107–31.

Schmidt, K.P. (1954). Faunal realms, regions and provinces. *Quarterly Review of Biology*, **29**, 322–36.

Schoener, A. (1988). Experimental island biogeography. In *Analytical biogeography* (ed. A.A. Myers and P.S. Giller), pp. 483–512. Chapman and Hall, London.

Sclater, J.G., Hellinger, S., and Tapscott, C. (1977). The paleobathymetry of the Atlantic Ocean from the Jurassic to the present. *Journal of Geology*, **85**, 509–52.

Sclater, P.L. (1858). On the general geographical distribution of the members of the Class Aves. *Journal and Proceedings of the Linnean Society of London (Zoology)*, **2**, 130–60.

Scotese, C.R. and McKerrow, W.S. (1990). Revised maps and introduction. In *Palaeozoic, palaeogeography and biogeography*, Geological Society. Memoir no. 12.(ed. W.S. McKerrow and C.R. Scotese), pp. 1–21.

Scotese, C.R., Gahagan, L.M., and Larsen, R.L. (1988). Plate tectonic reconstructions of the Cretaceous and Cenozoic ocean basins. *Tectonophysics*, **155**, 27–48.

Segerstråle, S.G. (1957). Baltic Sea. *Geological Society of America Memoir*, **67**, 751–800.

Seilacher, A. (1967). Bathymetry of trace fossils. *Marine Geology*, 5, 413–28.

Sengör, A.M.C. (1984). The Cimmeridge orogenic system and the tectonics of Eurasia. *Geological Society of America Special Paper 195.*

Sengör, A.M.C. and Yilmaz, Y. (1981). Tethyan evolution of Turkey: a plate tectonic approach. *Tectonophysics*, 75, 181–241.

Sengör, A.M.C., Altiner, D., Cin, A., Ustaömer, T., and Hsü, K.J. (1988). Origin and assembly of the Tethyside orogenic collage at the expense of Gondwana Land. In *Gondwana and Tethys* (ed. M.G. Audley–Charles and A. Hallam), pp. 119–81. Oxford University Press, Oxford.

Senowbari–Daryan, B. and Stanley, G.D. (1986). Thalassinid anomuran microcoprolites from Upper Triassic carbonate rocks of Central Peru. *Lethaia*, 19, 343–54.

Sepkoski, J.J. (1981). A factor analytic description of the Phanerozoic marine fossil record. *Paleobiology*, 7, 36–53.

Sepkoski, J.J. (1986). Phanerozoic review of mass extinction. In *Patterns and processes in the history of life.* (ed. D.M. Raup and D. Jablowski), pp. 259–76. Springer Verlag, Berlin.

Sepkoski, J.J. (1991). A model of onshore–offshore change in faunal diversity. *Paleobiology*, 17, 58–77.

Sepkoski, J.J., Bambach, R.K., Raup, D.M., and Valentine, J.W. (1981). Phanerozoic marine diversity and the fossil record. *Nature*, 293, 435–7.

Shackleton, N.J. (1984). Oxygen isotope evidence for Cenozoic climatic change. In *Fossils and Climate* (ed. P.J. Brenchley), pp. 27–34. Wiley, Chichester.

Shackleton, N.J. and Kennett, J.P. (1975). Paleotemperature history of the Cenozoic and the initiation of Antarctic glaciation: oxygen and carbon isotope analysis in D.S.B.P. sites 277, 279 and 281. *Initial Reports of the Deep Sea Drilling Project*, 29, 743–55.

Sheehan, P.M. and Coorough, P.J. (1990). Brachiopod zoogeography across the Ordovician–Silurian extinction event. In *Palaeozoic, palaeogeography and biogeography. Geological Society Memoir no. 12. (ed. W.S. McKerrow and C.R. Scotese), pp. 181–7.*

Shubin, N.H. and Sues, H.D. (1991). Biogeography of early Mesozoic continental tetrapods : patterns and implications. *Paleobiology*, 17, 214–30.

Silberling, N.J. (1985). Biogeographic significance of the Upper Triassic bivalve *Monotis* in circum-Pacific accreted terranes. In *Tectonostratigraphic terranes of the circum-pacific region* (ed. D.G. Howell), pp. 63–70. Circum-Pacific Council for Energy and Mineral Resources, Houston. Earth Science Series no. 1.

Simpson, G.G. (1940). Mammals and land bridges. *Journal of the Washington Academy of Sciences*, 30, 137–63.

Simpson, G.G. (1943). Mammals and the nature of continents. *American Journal of Science*, 241, 1–31.

Simpson, G.G. (1947). Holarctic mammalian faunas and continental relationships during the Cenozoic. *Bulletin of the Geological Society of America* 58, 613–88.

Simpson, G.G. (1950). History of the fauna of Latin America, *American Scientist*, 38, 261–89.

Simpson, G.G. (1952). Probabilities of dispersal in geological time. *Bulletin of the American Museum of Natural History*, 99, 163–76.

Simpson, G.G. (1980). *Splendid isolation: the curious history of South American mammals*. Yale University Press, New Haven, Connecticut.

Skelton, P.W. (1988). The trans-Pacific spread of equatorial shallow-marine benthos in the Cretaceous. In *Gondwana and Tethys* (ed. M.G. Audley-Charles and A. Hallam), pp. 247–53. Oxford University Press, Oxford.

Skevington, D. (1974). Controls influencing the composition and distribution of Ordovician graptolite faunal provinces. *Special Papers in Palaeontology*, **13**, 59–73.

Smelror, M. (1993). Biogeography of Bathonian to Oxfordian (Jurassic) dinoflagellates: Arctic, NW Europe and circum-Mediterranean regions. *Palaeogeography, Palaeoclimatology, Palaeoecology*, **102**, 121–60.

Smith, A.B. (1988). Late Palaeozoic biogeography of East Asia and palaeontological constraints on plate tectonic reconstructions. *Philosophical Transactions of the Royal Society of London*, **A 326**, 189–227.

Smith, A.B. (1992). Echinoid distribution in the Cenomanian: an analytical study in biogeography. *Palaeogeography, Palaeoclimatology, Palaeoecology*, **92**, 263–76.

Smith, A.G. and Briden, J.C. (1977). *Mesozoic and Cenozoic palaeocontinental maps. Cambridge University Press, Cambridge.*

Smith, A.G. and Livermore, R.A. (1991). Panagea in Permian to Jurassic time. *Tectonophysics* **187**, 135–79.

Smith, A.G., Briden, J.C., and Drewry, G.E. (1973). Phanerozoic world maps. *Special Papers in Palaeontology*, **12**, 1–42.

Smith, A.G., Hurley, A.M. and Briden, J.C. (1981). Phanerozoic palaeocontinental maps. Cambridge University Press, Cambridge.

Smith, P.L. (1983). The Pliensbachian ammonite *Dayiceras dayiceroides* and Early Jurassic paleogeography. *Canadian Journal of Earth Sciences*, **20**, 86–91.

Smith, P.L. and Tipper, H.W. (1986). Plate tectonics and paleobiogeography: Early Jurassic (Pliensbachian) endemism and diversity. *Palaios*, **1**, 399–412.

Sohl, N.F. (1987). Cretaceous gastropods: contrasts between Tethys and the temperate provinces. *Journal of Paleontology*, **61**, 1085–111.

Sondaar, P.Y. (1977). Insularity and its effects on mammal evolution. In *Major patterns in vertebrate evolution* (ed. M.K. Hecht, P.C. Goody, and B.M. Hecht), pp. 671–707. Plenum Press, New York.

Spicer, R.A., Rees, P. McA., and Chapman, J.L. (1993). Cretaceous phytogeography and climatic signals. *Philosophical Transactions of the Royal Society of London*, **B 341**, 277–86.

Spörli, K.B. (1987). Development of the New Zealand microcontinent. In *Circum-Pacific orogenic belts and evolution of the Pacific Ocean Basin* (ed. J.W.H. Monger and J. Francheteau), pp. 115–32. Geodynamic Series 18. American Geophysical Union, Washington.

Spörli, K,B., Aita, Y., an Gibson, G.W. (1989). Juxtaposition of Tethyan and non-Tethyan Mesozoic radiolarian faunas in melanges, Waipapa terrane, North Island, New Zealand. *Geology*, **17** 753–56.

Srivastava, S.K. (1978). Cretaceous spore-pollen floras: a global evaluation. *Biological Memoirs*, **3**, 2–130.

Stanley, S.M. (1984). Marine mass extinction: a dominant role for temperature. In *Extinctions* (ed. M.H. Nitecki), pp. 69–117. University of Chicago Press, Chicago.

Stanley, S.M. (1986). Anatomy of a regional mass extinction: Plio-Pleistocene decimation of the western Atlantic bivalve fauna. *Palaios*, **1**, 17–36.

Stanley, S.M. (1987). *Extinction*. Scientific American Library, New York.

Stanley, S.M. (1992). An ecological theory for the origin of *Homo*. *Paleobiology*, **18**, 237–57.

Stebbins, G.L. (1974). *Flowering plants : evolution above the species level*. Harvard University Press, Cambridge, Massachusetts.

Stehli, F.G. (1968). Taxonomic diversity gradients in pole location : the Recent model. In *Evolution and environment* (ed. E.T. Drake), pp. 163–277. Yale University Press, New Haven, Connecticut.

Stehli, F.G. and Webb, S.D. (ed) (1985). *The great American biotic interchange*. Plenum Press, New York.

Stehli, F.G., McAlester, A.L., and Helsey, C.E. (1967). Taxonomic diversity of Recent bivalves and some implications for geology. *Bulletin of the Geological Society of America*, **78**, 455–66.

Stenseth, N.C. and Maynard Smith, J. (1984). Coevolution in ecosystems: Red Queen or stasis? *Evolution* **38**, 870–80.

Stevens, C.H. (1983). Corals from a dismembered late Paleozoic paleo-Pacific plateau. *Geology*, **11**, 603–6.

Stevens, G.R. (1973*a*). Jurassic belemnites. In *Atlas of palaeobiogeography* (ed. A. Hallam), pp. 259–73. Elsevier, Amsterdam.

Stevens, G.R. (1973*b*). Cretaceous belemnites. In *Atlas of palaeobiogeography* (ed. A. Hallam), pp. 385–401. Elsevier, Amsterdam.

Stevens, G.R. (1980). Southwest Pacific faunal biogeography in Mesozoic and Cenozoic times : a review. *Palaeogeography, Palaeoclimatology, Palaeoecology*, **31**, 153–96.

Stevens, G.R. (1990). The influence of palaeogeography, tectonism, and eustasy on faunal development in the Jurassic of New Zealand. In *Fossili, evoluzione, ambiente* (ed. G. Pallini *et al.*), pp. 441–57. Atti II Conv. Int. E.E.A. Pergola.

Streel, M., Fairon-Demaret, M., and Loboziak, S. (1990). Givetian–Frasnian phytogeography of Euramerica and western Gondwana based on miospore distribution. In *Palaeozoic palaeogeography and biogeography*, Geological Society Memoir no. 12. (ed. W.S. McKerrow and C.R. Scotese), pp. 291–6.

Sylvester-Bradley, P.C. (1971). Dynamic factors in animal palaeogeography. In *Faunal provinces in space and time* (ed. F.A. Middlemiss, P.F. Rawson, and G. Newall), pp. 1–18, Seel House Press, Liverpool.

Talent, J.A., Gratsianova, R.T., and Yolkin, E.A. (1987). Prototethys : fact or phantom? Palaeobiogeography in relation to the crustal mosaic for the Asia–Australia hemisphere in Devonian–Early Carboniferous time. In *Shallow Tethys 2* (ed. K.G. McKenzie), pp. 87–111. Balkema, Rotterdam.

Tarling, D.H. and Runcorn, S.K. (1973). *Implications of continental drift to the earth sciences*, vol. 1. Academic Press, London.

Taylor, D.G., Callomon, J.H., Hall, R., Smith, P.L., Tipper, H.W., and Westermann, G.E.G. (1984). Jurassic ammonite biogeography of western North America : the tectonic implications. *Geological Association of Canada Special Paper*, **27**, 121–42.

Taylor, P.D. and Larwood, G.P. (ed.) (1990). *Major evolutionary radiations*. Oxford University Press, Oxford.

Tedford, R.H. (1974). Marsupials and the new paleogeography. In *Paleobiogeography* (ed. C.A. Ross). Benchmark Papers in Geology 31, pp. 109–26. Dowden, Hutchinson and Ross, Stroudsburg, Pennsylvania.

Thiede, J. (1977). Subsidence of aseismic ridges : evidence from sediments on Rio

Grande Rise (south-west Atlantic Ocean). *Bulletin of the American Association of Petroleum Geologists*, **61**, 939–40.

Thierry, J. (1976). Paléobiogéographie de quelques Stephanocerataceae (Ammonitina) du Jurassique moyen et supérieur : une confrontation avec la théorie mobiliste. *Geobios* **9**, 291–331.

Thierry, J. (1988). Structure and palaeogeography of the western Tethys during the Jurassic : tests based on ammonite palaeobiogeography. In *Gondwana and Tethys* (ed. M.G. Audley-Charles and A. Hallam), pp. 225–34. Oxford University Press, Oxford.

Thomas, H., (1984). Les Bovidae (Artiodactyla : Mammalia) du Miocène du sous-continent Indien, de la peninsule Arabique et de l'Afrique : biostratigraphie, biogeographie et ecologie. *Palaeogeography, Palaeoclimatology, Palaeoecologie*, **45**, 251–99.

Thomas, H., Bernor, R., and Jaeger, J.J. (1982). Origines du peuplement mammalien en Afrique du Nord durant le Miocène terminal. *Geobios*, **15**, 283–397.

Thorson, G. (1961). Length of pelagic larval life in marine bottom invertebrates as related to larval transport by ocean currents. In *Oceanography* (ed. M. Sears), pp. 455–74. American Association for the Advancement of Science, Washington.

Tollman, A. and Kristan-Tollman, E. (1985). Paleogeography of the European Tethys form Paleozoic to Mesozoic and the Triassic relations of the eastern part of Tethys and Panthalassa. In *The Tethys* (ed. K. Nakazawa and J.M. Dickins), pp. 3–22. Tokai University Press, Tokyo.

Torsvik, T.H., Smethurst, M.A., Van der Voo, R., Trench, A., Abrahamsen, N., and Halvorsen, E. (1992). Baltica. A synopsis of Vendian–Permian palaeomagnetic data and their palaeotectonic implications. *Earth Science Reviews*, **33**, 133–52.

Tozer, E.T. (1982). Marine Triassic faunas of North America : their significance for assessing plate and terrane movements. *Geologische Rundschau*, **71**, 1077–104.

Traverse, A. (1988). Plant evolution dances to a different beat. Plant and animal evolutionary mechanisms compared. *Historical Biology*, **1**, 277–301.

Tuckey, M.E. (1990). Distributions and extinctions of Silurian Bryozoa. In Palaeozoic, palaeogeography and biogeography, Geological Society Memoir no. 12. (ed. W.S. McKerrow and C.R. Scotese), pp. 197–206.

Turner, S. and Tarling, D.H. (1982). Thelodont and other agnathan distributions as tests of Lower *Palaeozoic continental reconstructions. Palaeogeography, Palaeoclimatology, Palaeoecology*, **39**, 295–311.

Tyson, R.V. and Pearson, T.H. (ed) (1991). *Modern and ancient continental shelf anoxia*. Geological Society Special Publication no. 58.

Vakhrameev, V.A. (1991). *Jurassic and Cretaceous floras and climates of the earth*. Cambridge University Press, Cambridge.

Valentine, J.W. (1967). Influence of climatic fluctuations on species diversity within the Tethyan provincial system. In *Aspects of Tethyan biogeography* (ed. C.G. Adams and D.V. Ager). Systematics Association Publication 7, pp. 153–66.

Valentine, J.W. (1973). *Evolutionary paleoecology of the marine biosphere*. Prentice-Hall, Englewood Cliffs, New Jersey.

Valentine, J.W. (1984). Neogene marine climate trends : implications for biogeography and evolution of the shallow-sea biota. *Geology* **12**, 647–50.

Valentine, J.W. (ed.) (1985). *Phanerozoic diversity patterns*. Princeton University Press, Princeton.

Valentine, J.W., Foin, T.C., and Peart, D. (1978). A provincial model of Phanerozoic marine diversity. *Paleobiology*, **4**, 55–66.

Van der Voo, R. (1988). Paleozoic paleogeography of North America, Gondwana, and displaced terranes : comparisons of paleomagnetism with paleoclimatology and biogeographical patterns. *Bulletin of the Geological Society of America*, **100**, 311–24.

Vannier, J.M.C., Siveter, D.J., and Schallreuter, R.E.L. (1989). The composition and palaeogeographical significance of the Ordovician ostracode faunas of southern Britain, Baltoscandia, and Ibero-Armorica. *Palaeontology*, **32**, 163–222.

Van Valen, L. (1973). A new evolutionary law. *Evolutionary Theory*, **1**, 1–30.

Van Zinderen Bakker, E.M. and Coetzee, J.A. (1988) A review of Late Quaternary pollen studies in East, Central and Southern Africa. *Reviews of Palaeobotany and Palynology*, **55**, 73–81.

Van Zinderen Bakker, E.M. and Mercer, J.H. (1986). Major late Cainozoic climatic events and palaeoenvironmental changes in Africa viewed in a worldwide context. *Palaeogeography, Palaeoclimatology, Palaeoecology*, **56**, 217–35.

Vartanyan, S.L., Garratt, V.E., and Sher, A.V. (1993). Holocene dwarf mammoths from Wrangel Island in the Siberian Arctic. *Nature*, **362**, 337–40.

Veevers, J.J. (1988). Gondwana facies started when Gondwanaland merged in Pangaea. *Geology*, **16**, 732–4.

Veevers, J.J. and Powell, C.McA. (1987). Late Paleozoic glacial episodes in Gonwanaland reflected in transgressive–regressive depositional sequences in Euramerica. *Bulletin of the Geological Society of America*, **98**, 475–87.

Veevers, J.J., Powell, C.McA., and Johnson, B.D. (1980). Seafloor constraints on the reconstruction of Gonwanaland. *Earth and Planetary Science Letters*, **51**, 435–44.

Vermeij, G.J. (1977). The Mesozoic marine revolution : evidence from snails, predators and grazers. *Paleobiology*, **3**, 245–58.

Vermeji, G.J. (1978). *Biogeography and adaptation : patterns of marine life.* Harvard University Press, Cambridge.

Vermeji, G.J. (1986). Survival during biotic crises : the properties and evolutionary significance of refuges. In *Dynamics of extinction* (ed. D.K. Elliott), pp. 231–46. Wiley, New York.

Vermeji, G.J. (1987). *Evolution and escalation.* Princeton University Press, Princeton.

Vermeji, G.J. (1991). Anatomy of an invasion : the trans-Arctic interchange. *Paleobiology*, **17**, 281–307.

Vermeij, G.J. (1992). Trans-equatorial connections between biotas in the temperate eastern Atlantic. *Marine Biology*, **112**, 343–8.

Vishnevskaya, V. (1992). Significance of Mesozoic radiolarians for tectono-stratigraphy in Pacific rim terranes of the former USSR. *Palaeogeography, Palaeoclimatology, Palaeoecology*, **96**, 23–39.

Wallace, A.R. (1876). *The geographic distribution of animals.* Macmillan, London.

Wang, Y., Boucot, A.J., Rong, J.Y., and Yang, X.C. (1984). Silurian and Devonian biogeography of China. *Bulletin of the Geological Society of America*, **95**, 265–79.

Waterhouse, J.B. and Bonham-Carter, G.F. (1975). Global distribution and character of Permian biomes based on brachiopod assemblages. *Canadian Journal of Earth Sciences*, **12**, 1085–146.

Webb, S.D. (1985). Late Cenozoic mammal dispersals between the Americas. In

The great American biotic interchange (ed. F.G. Stehli and S.D. Webb), pp. 351–86. Plenum Press, New York.

Webb, S.D. (1991). Ecogeography and the Great American Interchange. *Paleobiology*, **17**, 266–80.

Webb, S.D. and Barnosky, A.D. (1989). Faunal dynamics of Pleistocene mammals. *Annual Reviews of Earth and Planetary Sciences*, **17**, 413–38.

Wegener, A. (1924). *The origin of continents and oceans*. Methuen, London.

Weijermars, R. (1986). Slow but not fast global expansion may explain the surface dichotomy of Earh. *Physics of the Earth and Planetary Interiors*, **43**, 67–89.

Weissel, J.K., Hayes, D.E., and Herron, E.M. (1977). Plate tectonics synthesis: the displacements between Australia, New Zealand and Antarctica since the late Cretaceous. *Marine Geology*, **25**, 231–77.

Wen, S.X. (1986). Jurassic bivalve faunas of the Qinghai-Xizang (Tibet) Plateau in West China and their palaeobiogeography. In *Shallow Tethys 2* (ed. K.G. McKenzie), pp. 247–52. Balkema, Rotterdam.

Wesley, A. (1973). Jurassic plants. In *Atlas of palaeobiogeography* (ed. A. Hallam), pp. 329–37. Elsevier, Amsterdam.

Westermann, G.E.G. (1973). The late Triassic bivalve *Monotis*. In *Atlas of palaeobiogeography* (ed. A. Hallam), pp. 251–7. Elsevier, Amsterdam.

Westermann, G.E.G. (1981). Ammonite biochronology and biogeography of the circum-Pacific Middle Triassic. In *The Ammonoidea* (ed. M.R. House and J.R. Senior), pp. 459–98. Academic Press, London.

Westermann, G.E.G. (1988). Middle Jurassic ammonite biogeography supports ambi-Tethyan origin of Tibet. In *Gondwana and Tethys* (ed. M.G. Audley-Charles and A. Hallam), pp. 235–9. Oxford University Press, Oxford.

Westermann, G.E.G. (ed.) (1992). *The Jurassic of the Circum-Pacific*. Cambridge University Press, Cambridge.

Westermann, G.E.G. (1993). Global bio-events in mid-Jurassic ammonites controlled by seaways. In *The Ammonoidea : environment, ecology, and evolutionary change* (ed. M.R. House), pp. 187–226. Oxford University Press, Oxford.

Whewell, W. (1840). *The philosophy of the inductive sciences*. Parker, London.

Whittington, H.B. and Hughes, C.P. (1972). Ordovician geography and faunal provinces deduced from trilobite distribution. *Philosophical Transactions of the Royal Society of London*, **B263**, 235–78.

Whitmore, T.C., (ed.) (1981a). *Wallace's line and plate tectonics*. Clarendon Press, Oxford.

Whitmore, T.C. (1981b). Palaeoclimate and vegetation history. In *Wallace's line and plate Tectonics* (ed. T.C. Whitmore), pp. 36–42. Clarendon Press, Oxford.

Whitmore, T.C. and Prance, G.T. (ed.) (1987). *Biogeography and Quaternary history in tropical America*. Clarendon Press, Oxford.

Wieczorek, J. (1988). Biogeography of Tithonian nerinacean gastropods. *Mémoires de la Société géologique de France*, 154, 35–9.

Wignall, P.B. (1993). Distinguishing between oxygen and substrate control in fossil benthic assemblages. *Journal of the Geological Society*, **150**, 193–6.

Williams, A. (1973). Distribution of brachiopod assemblages in relation to Ordovician palaeogeography. *Special Papers in Palaeontology*, **12**, 1–42.

Williamson, M. (1988). Relationship of species number to area, distance and other variables. In *Analytical biogeography* (ed. A.M. Myers and P.S. Giller), pp. 91–146. Chapman and Hall, London.

Wilson, J.T. (1966). Did the Atlantic close and then re-open? *Nature*, **211**, 676.

Winterer, E.L. (1991). The Tethyan Pacific during Late Jurassic and Cretaceous times. *Palaeogeography, Palaeoclimatology, Palaeoecology*, **87**, 253–65.

Wise,, S.W. (1988). Mesozoic-Conozoic history of calcareous nannofossils in the region of the Southern Ocean. *Palaeogeography, Palaeoclimatolohgy, Palaeoecology*, **67**, 157–79.

Witzke, B.J., Frest, T.J., and Strimple, H.L (1979). Biogeography of the Silurian – Lower Devonian echinoderms. (In *Historical biogeography, plate tectonics and the changing environment* (ed. J. Gray and A.J. Boucot), pp. 117–29. Oregon State University Press, Corvallis.

Wolfe, J.A. (1978). A paleobotanical interpretation of Tertiary climates in the northern hemisphere. *American Scientist*, **66**, 694–703.

Wolfe, J.A. (1980). Tertiary climates and floristic relationships at high latitudes in the northern hemisphere. *Palaeogeography, Palaeoclimatology, Palaeoecology*, **30**, 313–23.

Wolfe, J.A. and Leopold, E.B. (1967). Neogene and Early Quaternary vegetation of northwestern North America and northeastern Asia. In *The Bering land bridge* (ed. D.M. Hopkins), pp. 193–206. Stanford University Press, Stanford.

Woodburne, M.O. and Zinsmeister, W.J. (1984). The first land mammal from Antarctica and its biogeographic implications. *Journal of Paleontology*, **58**, 913–48.

Wyss, A.R., Flynn., J.J., Novell, M.A., Swisher, C.C., Charrier, R., Novacek, M.J., and McKenna, M.C. (1993). South America's earliest rodent and recognition of a new interval of mammalian evolution. *Nature*, **365**, 434–7.

Yancey, T.E. (1979). Permian positions of the northern hemisphere continents as determined from marine biotic provinces. In *Historical biogeography, plate tectonics and the changing environment* (ed. J. Gray and A.J. Boucot), pp. 239–47. Oregon State University Press, Corvallis.

Yemane, K., Bonnefille, R., and Faure, H. (1985). Palaeoclimatic and tectonic implications of Neogene microflora from the northwestern Ethiopian highlands. *Nature*, **318**, 653–6.

Young, G.C. (1981). Biogeography of Devonian vertebrates. *Alcheringa*, **5**, 225–43.

Young, G.C. (1990). Devonian vertebrate distribution patterns and cladistic analysis of palaeogeographic hypotheses. In *Palaeozoic, palaeogeography and biogeography*, Geological Society Memoir no. 12. (ed.W.D. McKerrow and C.R. Scotese), pp. 243–255

Zeiss, A. (1968). Untersuchungen zur Paläontologie der Cephalopoden der Unter-Tithon der südlichen Frankenalb. *Bayerische Akademie der Wissenschaft, Mathematiker Wissenschaftliche Klasse Abhandlungen* N.S. **132**, 1–190.

Zhuravlev, A.Y. (1986). Evolution of archaeocyaths and palaeobiogeography of the early Cambrian. *Geological Magazine*, **123**, 377–385.

Ziegler, A.M. (1990). Phytogeographic patterns and continental configurations during the Permian Period. In *Palaeozoic, palaeogeography and biogeography*, Geological Society Memoir no. 12. (ed. W.S. McKerrow and C.R. Scotese), pp. 363–79.

Ziegler, A.M., Bambach, R.K., Parrish, J.T., Barrett, S.F., Gierlowski, E.H., Parker, W.C.*et al.* (1981). Paleozoic biogegraphy and climatology. In *Paleobotany, paleoecology, and evolution,*(ed. K.J. Niklas), pp. 231–66. Praeger, New York.

Ziegler, A.M., Hulver, M.L., Lottes, A.L., and Schmachtenberg, W.F. (1984).

Uniformitarianism and palaeoclimates: inferences from the distribution of carbonate rocks. In *Fossils and Climate* (ed. P.J. Brenchley), pp. 3–25. Wiley, Chichester.

Ziegler, A.M., Parrish, J.M., Y., Gyllenhaal, E.D., and Rowley, D.B. (1993). Early Mesozoic phytogeography and climate. *Philosophical Transactions of the Royal Society of London*, **B 341**, 297–305.

Ziegler, B. (1967). Ammoniten Ökologie am Beispiel des Oberjura. *Geologische Rundschau*, **56**, 439–64.

Ziegler, P.A. (1987). Evolution of the Arctic-North Atlantic and the western Tethys. *American Association of Petroleum Geologists Memoir*, **43**, 1–198

Zinsmeister, W.J. (1982). Late Cretaceous–early Tertiary molluscan biogeography of the southern circum-Pacific. *Journal of Paleontology*, **56**, 84–102.

INDEX